W9-AKB-132

REINVENTING MEDICINE

Also by Larry Dossey, M.D.

Be Careful What You Pray For . . . You Just Might Get It
Prayer Is Good Medicine
Healing Words: The Power of Prayer and the Practice of Medicine
Meaning and Medicine
Recovering the Soul
Beyond Illness
Space, Time & Medicine

REINVENTING MEDICINE

Beyond Mind-Body to a New Era of Healing

LARRY DOSSEY, M.D.

HarperSanFrancisco
A Division of HarperCollins*Publishers*

Grateful acknowledgment is made to *The Journal of Scientific Exploration,* P.O. Box 5848, Stanford, CA, phone (650) 539-8581, fax (650) 595-4466, email: sims@jse.com, www.jse.com, for permission to quote from *JSE* 10 (1), 3:30, © 1996; to Barbara Mathews, M.Ed., Director, Multisensory Learning, Inc. Customized Seminars and Keynotes; to Naomi Epel for permission to quote from *Writers Dreaming* by Naomi Epel, © 1993; to Mona Lisa Schulz, M.D. Ph.D., for permission to quote from *Awakening Intuition* by Mona Lisa Schulz, © 1998.

FIRST EDITION

Library of Congress Cataloging-in-Publication Data

Dossey, Larry
Reinventing medicine : beyond mind-body to a new era of healing / Larry Dossey.—1st ed.
Includes bibliographical references
ISBN 0–06–251622–1 (cloth)
ISBN 0–06–251644–2 (pbk.)
1. Medicine and psychology. 2. Mind and body. 3. Medicine, Psychosomatic. I. Title.
R726.5.D673 1999
610—dc21 99–27970

99 00 01 02 03 RRD(H) 10 9 8 7 6 5 4 3 2 1

For Bet and Garry

There I beheld the emblem of a mind
That feeds upon infinity. . . .

WILLIAM WORDSWORTH
The Prelude

CONTENTS

ACKNOWLEDGMENTS

Every one of my writings has been furnished to me by a thousand different persons, a thousand different things.

GOETHE

The premise of this book is that the mind is infinite. This means that my mind touches and is touched by those of everyone else, and that all minds are linked together. This creates challenges in acknowledging the sources of one's ideas, because their possible origins include everyone who has a mind.

Who, then, should I thank for encouragement and support? The list could be infinite, like the mind itself. Perhaps thanks to a few people will spread to all others via the One Mind of which we are all a part.

I am indebted to the people who have freely shared their experiences of illness, without whom this book would not exist. And to my colleagues in the trenches—the physicians and nurses, the scholars and clinicians—who are researching and applying the principles I discuss, I am immensely thankful.

I am grateful to my agents Arielle Eckstut, Kitty Farmer, and James Levine. Their support has been unstinting.

Elizabeth Perle, my editor at HarperSanFrancisco, helped me refine and shape these ideas. Her grasp of the concept of nonlocal mind is so profound that at first I was taken aback. A writer always expects technical expertise from his editor; when he receives penetrating insight into his subject in addition, he is blessed.

The more I have explored my nonlocal connections with the world, the more important my local grounding, my roots, have become. My parents gave me lasting lessons in love, which bind us together beyond this life. In fact, my mother was the source of my favorite, briefest, and most resoundingly positive book review: "If my boy wrote it, it's right!" My sister and brother, to whom this book is dedicated, remain pillars of support.

My deepest gratitude is to Barbara—author, nurse, healer, wife. She does not write about nonlocality; she lives it.

—LARRY DOSSEY, M.D.

REINVENTING MEDICINE

INTRODUCTION

Where does one person end and another person begin?

IRIS MURDOCH

During my first year of medical practice, I had a dream that shook my world.

In it Justin, the four-year-old son of one of my physician colleagues, was lying on his back on a table in a sterile exam room. A white-coated technician tried to place some sort of medical apparatus on his head. Justin went berserk—yelling, fighting, trying to remove the gadget in spite of the technician's persistent efforts. At the head of the table stood one of Justin's parents, trying to lend support. Time after time the technician tried to accomplish her task, but Justin would not relent. Exasperated, the technician finally gave up, abandoned the procedure, and walked away.

I awoke in the gray dawn with the sensation that the dream was the most vivid I had ever experienced—numinous, profound, "realer than real." In view of the dream's seemingly trivial content, I could not explain why I felt so deeply moved. I thought about waking my wife and telling her about it but decided not to disturb her. What sense would it make to her? We hardly knew Justin, having seen him only three or four times before.

It was time for me to dress and go to the hospital. The morning was busy, and I forgot about the dream until midday. Then, while I was lunching in the staff area with Justin's father, his wife entered the room holding Justin in her arms. The boy was visibly upset, with ruffled hair and tears streaming down his face. Justin and his mom had just come from the electroencephalography (EEG) laboratory, where the technician had tried to perform a brain wave test on the youngster. Her specialty was obtaining EEG tracings on children, and she considered herself the best in the business. Her record was flawless; never had she failed to obtain a quality recording—until she met Justin. After relating to her husband what had happened, Justin's mother left with the disconsolate boy in her arms. Her husband accompanied them out of the room and went to his office.

By this time my dream was replaying itself in my mind, and I was stunned. I had dreamed the sequence of events in exact detail before they happened. Disturbed, I went to see Justin's father in his office and asked him to tell me what had happened the day before.

Justin, he related, had developed a fever, which was followed by a brief seizure. Although he was certain the seizure was due to the fever and not to a serious condition such as epilepsy or a brain tumor, he nonetheless called a neurologist for a quick consultation. The specialist was reassuring; nothing needed to be done immediately. He would arrange an appointment for the following day for Justin to have a brain wave test at the EEG lab just to make sure nothing else was going on. It was a simple procedure, and the EEG technician, he said, had a special way with kids.

Could anyone else have known about these events? I asked. I wanted to know if someone could have leaked information to me that could have influenced my dream. Of course not, Justin's dad said; no one knew about the problem except the immediate family and the neurologist.

Then I told my colleague about my dream. He realized in an instant that if my report was true, his orderly, predictable world had been suddenly rearranged. If one could know the future before it happened, our understanding of physical reality was seriously threatened. He sensed my disturbance, and I sensed his. Our conversation dissipated into silence as we contemplated the implications of these events.

I turned, left his office, and closed the door behind me. I did not bring up the event with him again.

Within a week I dreamed two more times about events that occurred the next day, and that I could not possibly have known about ahead of time. It was as if the world had suddenly decided to reveal a new side of itself, for reasons I could not fathom.

In all three instances time seemed to be reversed, with effects appearing before their causes. Rationally I knew this was impossible. Time simply could not flow backward, carrying information into the present from a future that had not yet happened. I wondered, could my mind have strayed from my body and ventured into the future, retrieving knowledge about events that would occur later? Both possibilities violated every ounce of my medical training and apparent common sense. My consciousness was localized to my brain; *all* doctors knew that.

Those three dreams were the extent of my "future dreams." I had not had such a dream before, nor have I had one since. It was as if the universe, having delivered a message, hung up the phone. It was now up to me to make sense of it.

In the months that followed my dreams of the future, I resolved to learn more about dreams. I read several books on dream research and quickly discovered there was nothing special about my experience. I found, in fact, that precognitive, prophetic, or future dreams are quite common. Of great interest to me was the role dreams have played in healing throughout history. Early Greek physicians actually encouraged dreaming. When they treated their patients in their *asklepions,* temples devoted to healing, they instructed their patients to pay particular attention to dream content, because valuable therapeutic insights or actual healing often came during a dream. An example is Hippocrates (469–399 B.C.E.), who is regarded as the father of Greek medicine. He believed in prophetic dreams, diagnostic dreams, and dreams that revealed psychologically important messages. He theorized that during the day the sense organs are dominant and the soul is passive; but during sleep the emphasis shifts, and the soul then produces impressions instead of receiving them. Plato (427–347 B.C.E.) was interested in the emotional side of dreams. In the ninth book of the *Republic* he observed, "In all of us, even in good men, there is a lawless wild beast nature which peers out in sleep." Reasoning is suspended in sleep, he believed, which permitts the passions to reveal themselves in full force in dreams of murder, incest, and sacrilege. Aristotle (384–322 B.C.E.), Plato's student, suggested that dreams could be sensitive indicators of bodily conditions. During sleep, when external stimuli are minimized, the dreamer can pay closer attention to delicate internal sensations that might portend illness. The Greek physician Galen (ca. 130–ca. 200 C.E.) had a great impact on European medicine. A dream at age seventeen prompted him to shift his career from philosophy to medicine. He valued dreams in the diagnosis of illness, and he trusted dreams sufficiently to carry out surgical operations based on them. He claimed that he had saved many lives as a result. In contrast, the Roman orator Cicero (106–43 B.C.E.) was a great cynic about dreams. He believed the same dream could yield a variety of interpretations, and he held dream interpreters in disdain. "Let us reject, therefore, this divination of dreams, as well as all other kinds," he urged. "For, to speak truly, that superstition has extended itself through all nations, and has oppressed the intellectual energies of all men, and has betrayed them into endless imbecilities."[1]

These Greek and Roman views of dreams reflect their belief that the human mind is restricted to the individual body. The Persians, the neighbors of the Greeks, took a bolder view. They held that the mind could escape the confines of the physical body and create effects in the outside world. As the legendary Persian physician Avicenna (980–1037 C.E.) put it, "The imagination of man can act not only on his own body but even on others and very distant bodies. It can fascinate and modify them; make them ill, or restore them to health."[2]

One offshoot of classical Greek thinking held a place for the idea that the mind is both limited and infinite, human and divine. This was the lore surrounding the mythical sage Hermes Trismegistus. *Hermes* was the Greek name for the Egyptian god Thoth, believed to be the founder of alchemy and other occult sciences. The *Corpus Hermeticum,* a compendium of Greek and Latin writings dating to about two thousand years ago, is ascribed to Hermes. These writings are structured as private, intimate talks between a teacher and one or two of his disciples. In them we find the following conversation, in which the mythical Hermes is instructing the mythical Asclepius about the nature of humankind:

> Man is a marvel then, Asclepius; honour and reverence to such a being! . . . He is linked to the gods, inasmuch as there is in him a divinity akin to theirs. . . . He has access to all; he descends to the depths of the sea by the keenness of his thought; and heaven is not found too high for him. . . . Man is all things; man is everywhere.[3]

And in another discourse Hermes says,

> There is nothing more divine than mind, nothing more potent in its operation, nothing more apt to unite men to gods, and gods to men. . . . For man is a being of divine nature; he is comparable, not to the other living creatures upon earth, but to the gods in heaven . . . even above the gods of heaven. . . . None of the gods of heaven will ever quit heaven, and pass its boundary, and come down to earth; but man ascends even to heaven, and measures it . . . without quitting the earth; to so vast a distance can be put forth his power. . . . A man on earth is a mortal god, and . . . a god in heaven is an immortal man.[4]

But the Hermetic view of the infinite, divine mind was eventually eclipsed, and the aspect of Greek thought that came to dominate Western medicine was the earlier Hippocratic version—that the mind is a *local* event happening solely in the brain, in the here and now. The Persian and Hermetic views, which, in contrast, were *nonlocal*—that the mind is not localized or confined to the body but extends outside it—fell into neglect.

The dreams I experienced were Persian—a nocturnal magic carpet ride outside the here and now. I saw what lay ahead in great detail, three times running, which would not have been possible if the Hippocratic Greeks were right. Had I ventured into the future, or had the future drifted back into the present? And I didn't know *why* these events had happened. All I knew was that my consciousness seemed to be functioning outside of normal space and linear time, defying the way I'd been taught minds and brains behaved. But because of these experiences, I began to feel a new certainty—that I had touched something of infinite importance but something I was schooled to dismiss.

Yet the idea I touched was an ancient one, really. The possibility that the mind might function at a distance, outside the confines of the brain and body and not just in dreams, is taken for granted in most of what we call "native" cultures. David Unaipon, a native Australian, described in the early 1900s how the use of smoke signals depended on a nonlocal function of consciousness. Western travelers who witnessed this custom assumed that some sort of code was involved in the signaling. Not so, Unaipon explained; the function of the smoke signal was only to get everyone's attention so that distant, mind-to-mind communications might then take place:

> He might want to give his brother, who might be twenty miles away, a message; so he would set to and make a smoke signal, and then sit down and concentrate his mind on his brother. The column of smoke would be seen by all the blacks for miles around, and they would all concentrate their minds, and put their brains into a state of receptivity. Only his brother, however, would get in touch with him, and he . . . could then suggest to his [brother] the message which he wished to convey.[5]

For Native Americans, as for the Australian aboriginals, these nonlocal forms of communication were a crucial part of everyday life. They were used routinely in finding food and conducting warfare. Josiah Gregg, an authority on the American West during the heyday of the Oregon Trail, watched a Comanche arc an arrow that killed a prairie dog out of sight behind a hillock, a trick Gregg could not duplicate with a gun. When Victorio, the Apache leader, wished to know the location of the enemy, which often included the U.S. Cavalry, Lozen, his shaman, would stand with outstretched arms, palms up, and pray. As she turned slowly to follow the sun's path, her hands would begin to tingle and the palms change color when she faced the foe. The intensity of the sensation indicated the approximate distance to the enemy.[6]

We have repudiated these beliefs, yet my experiences tell me we have been too dismissive. Nor am I alone in my suspicion that we have stopped our investigation of healing well short of its potential. I have a file drawer full of letters from physicians around the world detailing nonlocal experiences they have had both in their personal lives and with their patients. I believe *every* physician becomes a collector of anomalies that break the rules—cures and remissions that don't fit the norm and that get quietly filed away over the years. Yet these experiences continue to inform our practice. Sometimes they come to us unbidden in dreams, as they did to me, perhaps because in dreams we drop our guard and stop censoring reality. When we release our prohibitions on what *can* happen, as we do while dreaming, we invite reality to frolic in its fullness.

The problem, of course, is what to think when we awaken. As Coleridge described this dilemma, "If a man could pass through Paradise in a dream, and have a flower presented to him as a pledge that his soul had really been there, and if he found that flower in his hand when he awoke—Aye! and what then?"[7]

THE ERAS OF MEDICINE

> *Modern consciousness research reveals that our psyches have no real or absolute boundaries; on the contrary, we are part of an infinite field of consciousness that encompasses all there is—beyond space-time and into realities we have yet to explore.*
>
> STANISLAV GROF

In so-called modern medicine, three distinct stages exist. The first one, which I call Era I, dates to the mid-1800s, when medicine first began to become a science—a method of practice supported by respected theories and undergirded by the accepted laws of nature. Scientists at that time regarded the entire world, including human bodies, as a mindless machine. Consciousness was increasingly seen as a material process that could be equated with the workings of the physical brain.

A hundred years later, however, the grip of this mechanical vision began to loosen. By the mid-1900s, physicians and scientists had begun to rediscover an old truth—that one's mind could affect one's body, sometimes dramatically. These insights came, for example, from research into the placebo response, in which expectations and positive thinking could alter the perception of pain as well as the response to various medications. Moreover, research revealed that illnesses such as hypertension and peptic ulcers developed when humans and animals were subjected to physical and psychological stress. As a result, a new epoch, Era II, began to take shape. Today this development is known as mind-body medicine.

Now, as the millennium turns, a new era, Era III, dawns. The hallmark of Era III is what I refer to as nonlocal mind. In Era III, we rediscover the ancient realization that consciousness can free itself from the body and that it has the potential to act not just *locally* on one's own body, as in Era II, but also *nonlocally* on distant things, events, and people, even though they may be unaware that they are being influenced.

There are those who resist Era III, and they usually do so because they cannot swallow the concept of nonlocal mind, which is the quintessential characteristic of Era III. Because of its nonmaterial properties, Era III feels strange and thus uncomfortable. In spite of scientific evidence, these people cling to a more familiar and mechanistic, materialistic view. They "know" that consciousness is local, that it stays put in the brain, in the body, and in the present moment. But just as mind-body medicine—which also aroused profound suspicion—has now gained popular acceptance, so, too, will nonlocal mind become critical to our understanding of healing ourselves and one another.

The two main lines of evidence for nonlocal mind are (1) the everyday experiences of millions of people and (2) scientific findings. In the chapters that follow we shall examine the lives of ordinary people and the work of scientists and physicians who embrace the premise of Era III medicine and who are translating its magic for the good of believers and skeptics alike.

WHY I USE THE TERM "NONLOCAL MIND"

I've spent years searching for a suitable term for the unbounded ways in which consciousness displays itself in space and time. I've bumped into dozens of expressions that have been used in the past—extended mind, cosmic consciousness, Mind at Large, the One Mind, universal mind, Buddha mind, Big Mind, Christ consciousness, collective consciousness, the collective unconscious, and on and on. One by one, I discarded all these terms. They seemed imprecise or dated or were burdened with too many religious or psychological associations. *Nonlocal* is a term used by physicists to describe the distant interactions of subatomic particles such as electrons. Since it has been proven that human minds display similar interactions at a distance, I chose to adopt the physicists' term—recognizing, however, that physics does not actually explain how nonlocal events happen at the human level. I first introduced it ten years ago in the book *Recovering the Soul.*[8] No other term, in my judgment, describes consciousness so well as *nonlocal mind,* for reasons that will emerge as we explore Era III.

WHY SHOULD WE PAY ATTENTION TO ERA III?

We have all had the experience of mental events that don't fit the commonsense pictures we have of reality. And we need a place to contain and harness the power of these occurrences. Some Era III "anomalies" include what are commonly called telepathy, clairvoyance, precognition, visions, prophetic dreams, instances of unexplained breakthroughs in discovery or creativity, the sharing of physical symptoms between distant individuals, distant healing, and intercessory prayer. Some researchers, seeking fresher, more scientific terms, have begun to call these events "anomalous cognition," "anomalous perturbation," "distant manifestations of consciousness," or the results of "distant intentionality." But no matter what term is chosen, the bottom line is the same: the mind is acting at a distance from the brain and body, and often outside the present moment.

Researchers are now beginning to pay attention to these rogue happenings, but they often do so in the face of tremendous resistance. Some fear they will be dismissed as fringe or New Age thinkers, others as advocates for God-inspired healing. Moreover, there is always pressure on scientists to dismiss these anomalous phenomena as irrelevant and confine their attention to "normal" science. "The scientist who pauses to examine every anomaly he notes," said the late historian of science Thomas Kuhn, "will seldom get significant work done." Even when we prove that an anomaly is real, "we do not usually surrender the intellectual system of a lifetime for one bit of information that does not fit," notes Stephen Jay Gould, the Harvard paleontologist. "Single facts almost never slay worldviews, at least not right away." Gould adds, "Anomalous facts are simply incorporated into existing theories, often with a bit of forced stretching to be sure, but usually with decent fit because most worldviews contain considerable flexibility."[9]

To be worth our time and attention, an anomaly must be more than just another puzzle of normal science. It must affect our personal, psychological, or spiritual life or our moral, political, and social situation. It must make a decisive impact; it must matter *a lot.* That is precisely what nonlocal mind does. All of us have personally experienced the nonlocal anomalies of consciousness, whether in thinking of someone and having that person call or, more dramatically, seeing a specific problem in our body or the body of a loved one before any warning signs or symptoms appeared. What we don't realize, though, is how close we are to discovering a new dimension of existence—a new era of being.

We need Era III. People everywhere are starved for meaning, purpose, and spiritual fulfillment in their lives. To lead healthy, full lives, we require a positive sense of meaning, just as we need food and water, for without meaning life withers. As we shall see, nonlocal mind is suffused with spiritual meaning and can help fill the inner void that has become such a painful feature of modern life.

I have come to these conclusions by painful personal experience. When I began my university education, I developed a profound love affair with science and the mechanical picture of reality of Era I. For many years thereafter, I tried to order my life according to this austere, rational vision. I loved this unadorned picture, because everything in it was regular and predictable, or so I thought. But the more I tried to fashion my life according to these principles, the more joyless it became. Moreover, ignoring my psychological and spiritual life led to an increase in the severity of a problem I'd had for years—classical migraine headaches with horrific pain, incapacitation, and blindness. This illness almost forced me to abandon my career in medicine, because I felt that it was only a matter of time until I experienced an attack of blindness while doing surgery and harmed or killed someone. In desperation, when all medical approaches had failed, I found that I could avoid these episodes by entering profound states of relaxation. I learned to do this through biofeedback training, which taught me to change my physiology through mental strategies. This discovery opened up for me a new world—the dimension of mind-body interaction of Era II—and led eventually to my entry into the spiritual dimensions of Era III.

I have a physician friend who frequently speaks to both lay audiences and groups of doctors about the role of the mind in health. I once asked him if he changes his talk when he addresses the physicians. "Yes," he said. "I make it simpler." How well I know the reason! What seems self-evident to laypeople is often inscrutable to us physicians. We are so committed to a mechanical view of reality that we typically deny evidence that challenges this vision. This is the reason I came, sometimes kicking and screaming, to the views discussed in this book.

Indeed, if there is anything unique about my personal journey, it is the high level of stubbornness that has characterized it.

I used to believe that we must choose between science and reason on the one hand and spirituality on the other, as foundations for living our lives. Now I consider this a false choice, because in my own life I have found that science and spirituality can coexist and even flourish. I have discovered that many great scientists have agreed with this point of view, such as Robert Boyle, the seventeenth-century chemist and author of Boyle's Law. Boyle called scientists priests of nature. He believed that science is so sacred that he recommended that scientists do their experiments on Sundays as part of their sabbath worship.

Era III makes it possible to recover the sense of sacredness Boyle recognized, not just in science but in every area of life. Let us explore how.

—LARRY DOSSEY, M.D.

CHAPTER 1

THE ERAS OF MEDICINE

Man's perceptions are not bounded by organs of perception: he perceives far more than sense (tho' ever so acute) can discover.

WILLIAM BLAKE

"snowing and about three inches deep . . . wind at northeast and mercury at 30. . . . Continuing snowing till one o'clock and about four it became perfectly clear. Wind in the same place but not hard. Mercury at 28 at night."[1]

These were the last words George Washington, the first president of the United States, wrote.

On the morning of December 13, 1799, at age sixty-seven, Washington had gone on his long daily ride at Mount Vernon. He was an obsessive horseman, and not even foul weather could keep him out of the saddle. When he returned later that day, his greatcoat was soaked through, and snow hung from his white hair. He sat down to dinner without changing his damp clothes, and by evening he had a sore throat. On trying to read parts of the newspaper aloud, he was hampered by hoarseness. When his secretary, Tobias Lear, suggested he take some medicine, Washington declined, saying, "No. You know I never take anything for a cold. Let it go as it came."

Between two and three in the morning, Washington woke his wife, Martha, and complained that he had a very sore throat and was feeling unwell. He could hardly talk, was shaking with chills, and had trouble breathing. At George's request, Martha sent for his lifelong friend Dr. William Craik, who had been his companion in the French and Indian War and a fellow explorer of the frontier. In the meantime, Washington asked Rawlins, the overseer who usually took care of sick slaves, to bleed him. He bared his arm, and Rawlins made the incision, but Washington complained that the incision was not wide enough. "More," he ordered. When Craik arrived he applied Spanish fly to Washington's throat, to draw blood into a blister, and bled him again. Washington was given sage tea and vinegar to gargle and nearly choked. Craik sent for another doctor and bled him again.

Between three and four in the afternoon, two more physicians, Gustavus Brown and Elisha Cullen Dick, arrived. Craik and Brown agreed on a diagnosis of quinsy, what we today would call acute streptococcal pharyngitis, or strep throat. They decided on more bleedings, blisterings, and purges with laxatives. But Dick, a thirty-seven-year-old graduate of the University of Edinburgh School of Medicine (also Craik's alma mater), dissented. It was his view that Washington was suffering from "a violent inflammation of the membranes of the throat, which it had almost closed, and which, if not immediately arrested, would result in death." Dick urged that a radical new surgical procedure be performed that he had learned about in Scotland for cases like this—a tracheotomy below Washington's infected, swollen throat to allow him to continue to breathe. But this was too much for the senior physicians Craik and Brown, and they would not agree.

Dick took another tack. At the very least, he pleaded, do not bleed Washington again. "He needs all his strength—bleeding will diminish it." Again Craik and Brown ignored the younger doctor. They asked for and obtained Washington's consent to bleed him a fourth time.

Washington rallied briefly, long enough for Craik to give him calomel and other purgatives.

Shortly thereafter, George asked Martha to come to his bedside. He requested that she bring his two wills and burn the old one, which she did.

Washington continued to defer to the advice of Craik and to refuse the suggestions of the younger man. He had convinced himself early in the day that he was going to die. "I find I am going. My breath cannot continue long," he whispered to Lear, to whom he gave instructions for the arrangements of all his military papers and accounts. Then Washington smiled and said with perfect resignation that death "is the debt which we must all pay." To Craik he whispered a little later, "Doctor, I die hard, but I am not afraid to go. My breath cannot last long." A lifelong stoic, he did not complain, although he must have been in terrific pain.

"'Tis well," he finally whispered. These were his last words. Five hours later, with his beloved Martha at his side, George Washington died.[2]

Washington was hard to kill. At a muscular six-four, he was a giant for his day. His ironlike constitution enabled him to survive a volley of illnesses that would have killed weaker men—dysentery, influenza, malaria, mumps, pleurisy, pneumonia, rickets, smallpox, staph infections, tuberculosis, and typhoid fever—not even counting all the lead shot at him. It is ironic that in the end he succumbed to an illness that today is regarded more as a nuisance than a disease and that can be cured by a single injection or a handful of pills: strep throat.[3]

It is easy to find fault with the way America's first president was treated in his final hours, but retrospective criticism is unfair. Washington's physicians were doing the best they could with the knowledge they had. To senior physicians Craik and Brown, young Dick's suggestion of a tracheotomy probably sounded like assassination. They were unwilling to make Washington, the most revered man in America, an experiment for an unproved, unfamiliar surgical intervention. Washington himself declined Dick's advice. A true man of his time, he got what he expected and what he wanted—bleeding, blistering, and purging.

THE ERAS OF MEDICINE

Washington's deathbed therapies show a gruesome side of medicine, which has prevailed for most of our Western history. His final hours reveal both the helplessness of the physicians of his day and the fact that by and large the techniques in use at the time either did not work or were actually harmful. In the early nineteenth century, there was no getting around the fact that doctors were dangerous. It has been proposed that the reason the kings and queens of Europe during this period lived shorter lives than ordinary folk, in spite of having adequate housing and nutrition, is that they had unlimited exposure to one of the greatest health risks of the day—the medical profession.[4]

It is difficult for us to comprehend the health risks people in Washington's time faced. If you were born in the United States at any time prior to the mid–nineteenth century, you had less than a 50 percent chance of surviving long enough to produce any children. It was particularly hazardous to live in a large city. Because the growth of cities following the Industrial Revolution occurred at a time when there were virtually no minimum standards or restrictions on building, the great cities in the Western world achieved a level of squalor and mortality almost matching that of ancient Rome. In the mid-1800s, one-fifth of all infants born in New York City died before reaching their first birthday, often of infant diarrhea, and even if they reached adulthood they had a one-in-four chance of succumbing between the ages of twenty and thirty. As for life in the country, the semirural districts of England in 1860 had a mortality level not much different from Roman North Africa more than fifteen hundred years earlier.[5] In the pastoral Tidewater area of Virginia, where Washington lived at Mount Vernon, epidemics of cholera, yellow fever, and smallpox often raged.

We often think that the diseases our predecessors endured came in waves, such as the bubonic plague, which reduced the population of

Europe by perhaps two-thirds between 1320 and 1420.[6] Or the epidemic of yellow fever that killed more than six thousand of Philadelphia's thirty thousand residents in August 1793, forcing George and Martha Washington to flee the city, which served at the time as the nation's capital. Yet there were no disease-free intervals between the epidemics. Illnesses such as typhoid and pneumonia killed at a steady rate, year after year. Death stalked everyone, everywhere, all the time.

Although I practiced internal medicine in a large city for more than twenty years, I never saw a case of cholera, smallpox, yellow fever, or polio, and I might have had trouble recognizing them if I had. In contrast, death from infections was so prevalent in America's large cities in the eighteenth and nineteenth centuries that people became experts in recognizing infectious diseases, and they would have put modern physicians to shame. Nancy Tomes, professor of history at State University of New York at Stony Brook, writes that all the social classes of that time were familiar with "the blue skin and rice-water discharges of cholera and the high fever and rash that signaled typhoid. They could recognize," she writes, "the characteristic skin eruptions of smallpox, and the sore throat, strawberry tongue, and sunburn-like rash of scarlet fever." They could even, she says, "differentiate the coughs associated with whooping cough, pneumonia, and consumption. They knew too well the chronic diarrhea and wasting that indicated the 'summer complaint' and the labored respiration and blocked airways produced by diphtheria."[7]

But they could do little about them. In Washington's time and for decades following, people did not understand that the diseases that carried them away were related to microorganisms. The "germ theory of disease" did not come into use in the English-language medical literature until around 1870. In addition to threats from infectious diseases, people faced broken bones, cancer, metabolic diseases, endocrine disorders, cardiovascular ailments, and on and on. The result was that people were burdened by a sense of mystery, vulnerability, chronic dread, and resignation—the certainty that "this is the way things are and have always been." And they were right, for in the three million or so years humans have existed, there had not been much change in the pattern of mortality up to their time.[8] And, following Washington's demise, things were not to change much in Western medicine for the next half-century.

Then, beginning roughly in the decade of the 1860s, the decade of the American Civil War, medicine began to change radically and dramatically by becoming scientific.

If we begin at that moment in history and move forward to the present day, we can sort out three distinct ways in which health, healing, and the nature of the mind have been viewed. Because these perspectives developed in a historical sequence, they can be called eras. Although these developments are referred to as *medical* eras, they reflect currents of thought that extend far beyond medicine. Medicine grows out of a culture; it is never isolated and self-contained. The following eras of medicine reflect, therefore, patterns that run deep in our culture.

As we examine these three eras, bear in mind that some of the ideas about the nature of the mind they contain are not new. In fact, the most recent view, that the mind operates outside the body, is, as we saw, quite ancient. But this idea is new to *us,* because we lost this understanding and are in the process of rediscovering it.

ERA I: MECHANICAL MEDICINE

Era I, the first scientific medical era, began to take shape in the last third of the nineteenth century, beginning in approximately the 1860s. Era I can be called materialistic, mechanistic, or physicalistic medicine. It encompasses the therapies that largely dominate Western medicine today—drugs, surgery, radiation, and so on. The classical laws of matter and energy described in the seventeenth century by Sir Isaac Newton form the foundation of Era I medicine. According to this view, the entire universe is a vast clockwork that functions according to deterministic, causal principles. In Era I, what matters is energy and matter, not mind.

MEDICAL ERAS

	ERA I	ERA II	ERA III
SPACE-TIME CHARACTERISTIC	Local	Local	Nonlocal
SYNONYM	Mechanical, material, or physical medicine	Mind-body medicine	Nonlocal, Era III, or Eternity Medicine
DESCRIPTION	Causal, deterministic, describable by classical concepts of space-time and matter-energy. Mind is not a factor; mind is a result of brain mechanisms.	Mind is a major factor in healing *within* the single person; mind has causal powers. Medicine is thus not fully explainable by classical concepts in physics. Era II includes but goes beyond Era I.	Mind is a factor in healing both *within* and *between* persons. Mind is not completely localized to points in space (brains or bodies) or time (present moment or single lifetimes). Mind is unbounded in space and time and thus is ultimately unitary or one. Healing at a distance is possible. Not describable by classical concepts of space-time or matter-energy.
EXAMPLES	Any form of therapy focusing solely on effects of *things* in the body. Almost all forms of modern medicine—drugs, surgery, irradiation, CPR, etc.	Any therapy emphasizing the effects of consciousness *solely* within the individual body: psychoneuroimmunology, counseling, hypnosis, biofeedback, relaxation therapies, and most of imagery-based "alternative" therapies.	Any therapy in which effects of consciousness bridge between different persons: all forms of distant healing, intercessory prayer, and transpersonal imagery.

Era I represented a monumental advance in the history of healing. It was fabulously effective, laying the groundwork for what we call conventional or orthodox medicine. In fact, the achievements of mechanical medicine are so significant that many people remain convinced that the future of medicine lies almost exclusively in Era I approaches. While much of this enthusiasm is justified, we have already moved beyond the confines of mechanistic medicine in our quest for health and healing, as we shall see.

I have spent most of my life as a physician working in Era I medicine, and I bow deeply in gratitude to its architects. I am grateful not just for the high-profile discoveries of Era I medicine, such as vaccines and antibiotics, but also for less spectacular measures, such as public health interventions and lifestyle modifications. These latter measures have been largely responsible for the improvement in health and longevity seen around the world in Era I, and they get less credit than they deserve. For example, half the reduction in mortality from diphtheria had already been achieved by public health measures by the time diphtheria immunization was introduced in the 1920s. And by the time antibiotic treatment became widely available in the 1940s, more than 90 percent of the mortality due to tuberculosis had already been prevented by other means.[9]

One of the most significant effects of Era I medicine was a revolution in how people *thought* about disease. Following the introduction of sulfonamides in the 1930s and penicillin in the 1940s, almost overnight the veil of helplessness began to lift, and victory over illness seemed possible. As John Cairns, formerly of the Harvard School of Public Health, describes the transformation, "Perhaps one of the main consequences . . . has been the idea that every disease should be treatable, if we could but find its Achilles' heel." Thus today we *assume* there should be a cure for any and every disease that afflicts us. Discovering it depends only on the availability of people, funding, and time. Prior to Era I medicine, people did not expect medicine to deliver cures. "In nineteenth-century London," Cairns points out, "there were riots about the failure of the government to prevent the spread of cholera but no complaints about the way the disease was being treated; 150 years later there are riots about the lack of any treatment for AIDS. . . . We are no longer content to endure illness and disease: the maimed and the halt expect to be healed, and healed quickly."[10]

Today most people place their confidence in high-tech solutions to health care. We are mesmerized by the promises of gene therapy, DNA manipulation, organ transplantation, computer-designed drugs, and so on. In this heady atmosphere it is often considered heresy to speak of limits to high-tech medicine. As Cairns says, "Most people, doctors especially, find it hard to believe that the main advances of medicine did not arrive until long after most of the causes of untimely death had come under control."[11] Era I medicine has always been limited and remains so. For example, it continues to be largely ineffective in combating chronic, degenerative, and age-related illnesses.

Acknowledging the limitations of Era I medicine is not disrespectful. Doing so opens us to other possibilities, other approaches.

ERA II: MIND-BODY MEDICINE

Following World War II, another discrete period in the history of Western medicine began to take shape, which I call Era II. This development grew out of research in the field of "psychosomatic disease"—from the Greek words *psyche,* meaning breath, spirit, or soul, and *soma,* meaning body.

The foreshadowings of Era II can be traced to antiquity. For fifty thousand years shamans have used expectation and suggestion to help people heal. The Greeks were well aware of the influence of the emotions on the body, and they honored the role of dreams in healing. Era I medicine, which took shape in the second half of the nineteenth century, was not devoid of this understanding, although most doctors left it decidedly in the background. Jean-Martin Charcot, the great nineteenth-century French neurologist, studied hysterical reactions and the effect of suggestion on bodily function. Sigmund Freud, a student of Charcot, extended these ideas and emphasized the influence of the unconscious on behavior. The end of World War I lent impetus to the idea that the mind could dramatically affect the body when thousands of soldiers returned home suffering from an emotional derangement called "shell shock," which was often incapacitating. But although the basic idea of mind-body interaction can be followed from ancient times forward, this premise did not affect the scientific medicine of Era I to any great extent until the period following World War II. Compared to the spectacular physical effects of Era I therapies such as antibiotics, vaccines, and irradiation, the mind's effects seemed trivial.

Moreover, by the middle of the twentieth century many authorities claimed there was simply no need for the concept of "mind" or "consciousness"; these terms were merely unnecessary symbols for the brain. But evidence began to mount suggesting that the mind, even if considered synonymous with the brain, could affect the body. The first effects to be noticed were negative ones—thus the term *psychosomatic disease.* For example, scientists demonstrated that if rats and mice were confined in close quarters or were stressed by exposure to electrical shocks, they developed gastrointestinal ulcerations, hypertension, and heart disease and often died. Humans, it was discovered, reacted similarly when under stress. Eventually it became clear that our emotions, attitudes, and thoughts profoundly affect our bodies, sometimes to the degree of life or death. Soon mind-body effects were recognized to have positive as well as negative impacts on the body. This realization came largely from research on the placebo effect—the beneficial results of suggestion, expectation, and positive thinking.

The mind-body approaches of Era II did not do away with the physical approaches of Era I. Rather, the two perspectives joined hands as allies. Today, for example, in treating cancer, psychological counseling is used *with* chemotherapy and surgery, not in place of them; for heart disease, stress management is used *with* medications and coronary bypass surgery, not as replacements for them.

Eras I and II may appear to have little in common, but they are quite similar in how they view the essential nature of the mind. In both, mind or consciousness is generally equated with the brain or is assumed to be a by-product or a result of the brain's chemistry and physiology—all this in spite of the fact that no one knows how the brain "makes" the mind, or even if it does. As astronomer Carl Sagan expressed this point of view, "[The brain's] workings—what we sometimes call mind—are a consequence of its anatomy and physiology, and nothing more."[12] Because most researchers equate mind and brain, *mind*-body research performed by them could more appropriately be called *brain*-body research.

We physicians often pay lip service to the importance of the mind in health and illness, yet in practice we equate it with the brain. This is obvious in our reliance on drugs when mental problems arise. For example, in the 1980s anxiety essentially came to be seen as a Valium deficiency in the brain, and in the nineties depression appears to be equated with a shortage of Prozac. Thus, Era II or "mind-body" medicine is often constrained by the purely physical "brain-body" medicine of Era I.

A refreshing contrast is alternative or complementary medicine, which has emerged as a major social force shaping medical care in the United States. In general, this field grants a major role for psychological and spiritual factors in health and does not consider the mind to be identical with the physical brain. Even when physical interventions such as acupuncture, homeopathy, or yoga are employed, alternative therapists generally recognize that the results of these therapies are affected by psychological and spiritual factors. A survey by John A. Astin, of Stanford University School of Medicine, shows that psychological and spiritual factors may be the main reasons people gravitate to alternative therapies. In his article "Why Patients Use Alternative Medicine" in the *Journal of the American Medical Association,* Astin says,

> Users of alternative health care are more likely to report having had a transformational experience that changed the way they saw the world. . . . They find in [alternative therapies] an acknowledgment of the importance of treating illness within a larger context of spirituality and life meaning. . . . The use of alternative care is part of a

> broader value orientation and set of cultural beliefs, one that embraces a holistic, spiritual orientation to life.[13]

In summary, Era I emphasized a completely physical, body-based approach to health and illness. Era II expanded the Era I focus by adding a place for the effects of the mind. Yet the mind-body effects of Era II remained centered within each individual—one's own mind affecting one's own body. This aspect of Era II is the limitation that opens to further expansion in Era III.

ERA III: NONLOCAL MEDICINE

Today, just as physicians are getting used to mind-body medicine, another great transition is taking shape. We are facing a "constitutional crisis" in medicine—a crisis over our *own* constitution, the nature of our mind and its relationship to our physical body. There are simply too many events that cannot be explained from the perspectives of Eras I and II to consider them a full explanation of reality. In order to encompass these phenomena and to advance healing, we are compelled to describe yet another era—Era III, or *nonlocal,* medicine.

Why "nonlocal"? In brief, nonlocal medicine does not confine or localize the mind to the brain and body. It grants the mind freedom to roam freely in space and time. In Era III, the mind is *more* than the brain; it can do things brains can't do, such as acting remotely from the body and venturing outside the present. As such, it can be used in ways not available in the mechanical methods of Era I or the mind-body approaches of Era II. But how?

Imagine that, as you read these words, a part of your mind is not present in your body or brain or even in this moment. Imagine that this aspect of your consciousness spreads everywhere, extending billions of miles into space, from the beginning of time into the limitless future, linking us with the minds of one another and with everyone who has ever lived or will live. This is the infinite piece of your consciousness.

This picture of your mind as outside your head may at first seem foreign. But as we shall see, "nonlocal" or "infinite" describes a natural part of who we are. Its expressions include sharing of thoughts and feelings at a distance, gaining information and wisdom through dreams and visions, knowing the future, radical breakthroughs in creativity and discovery, and many more. And this part of your mind can be used today in healing illness and disease in what I call Era III healing.

Many studies reveal that healing can be achieved at a distance by directing loving and compassionate thoughts, intentions, and prayers to others, who may even be unaware these efforts are being extended to them. These findings reveal the ability of some part of our mind or consciousness to escape its confinement to the brain and body and to act anywhere, regardless of distance. The medical implications of this are profound. For example, the mind's nonlocal nature provides a way for us to help heal one another, making health and illness a collective affair. Era III healing means that "we're in it together," that no one's illness is a purely individual matter. When illness occurs, the fact that I can help you and you can help me eases the isolation and feelings of being alone that are a painful part of being sick. Era III also provides new tools for physicians and health-care workers. If the mind operates beyond the body, this provides a way for physicians to know directly, intuitively, what is wrong with their patients. All physicians have had the experience of "just knowing" what a diagnosis is, with little or no information to go on. We attribute this to experience, but really it's more. It's intuition, too, and intuition is an important hallmark of the nonlocal mind.

But the greatest contribution of Era III to healing goes beyond diagnosing disease and eradicating illness. If our minds are genuinely capable of breaking free from the body, then they are not subject to time. The implication of this is so revolutionary, we instinctively dismiss it. Although our bodies die, the timeless part of our consciousness lives on. Era III provides a cure, therefore, for the "disease" that has caused more suffering for more people than any other—the fear of death.

Throughout history, the ability to extend one's mind beyond the body has often been considered a rare talent. And immortality, likewise, has often been granted to the spiritual elite—individuals who, through great dedication and effort, have achieved a rare state of existence. Yet we err when we assume that only an odd few possess these abilities. Nonlocal mind is not the possession of only a select group. Indeed it is so innate, *ordinary,* the gift belongs to everyone. Our minds are *naturally* nonlocal—present at birth, factory installed, freely given. This means that it isn't necessary for us to try to engineer nonlocal experiences. Because we *are* nonlocal, they occur without our bidding, often when we least expect them. We can learn, however, to set the stage for nonlocal events through various methods so that these happenings occur more frequently in our lives, as we shall see.

THREE TELLTALE SIGNS OF NONLOCAL HAPPENINGS

Since nonlocal mind is the defining characteristic of Era III, we need to examine it closely. *Nonlocal* does not mean merely "a long way off" or "a very long time" but, rather, *infinitude* in space and time. If something is nonlocal it is *unlimited.* Nonlocality, like pregnancy, is an all-or-nothing event. One cannot be "somewhat pregnant" or "a little nonlocal."

Recently, modern scientists have discovered that nonlocal events are not fantasy but are part of the fabric of the universe. Physicist and author Nick Herbert is an expert on nonlocality. He believes that it applies not just to the invisible world of atoms and electrons, but to the large-scale domain as well. He suggests that three features of nonlocality can be used to identify these events.

NONLOCAL HAPPENINGS ARE UNMEDIATED

Most of our experiences are "mediated"; that is, some sort of go-between makes them happen. For example, when we look at something, our visual experience is mediated by light traveling from the object seen to our retina. If we touch something, the experience is mediated by neural signals triggered by the pressure of the object on our hand. *All* sensory experiences are mediated. Not so with nonlocal experiences. Nonlocal events do not require a go-between signal. Indeed, researchers have diligently looked for some sort of subtle energy that connects distant individuals when thoughts are communicated at great distances or when prayer affects the body of someone far away. Yet there is not a shred of evidence that such energy exists.

NONLOCAL EVENTS ARE UNMITIGATED

This means that their strength does not "fall off" with increasing distance. If I throw a rock into San Francisco Bay, it makes a splash and creates a wave, but these effects cannot be detected in the Tokyo harbor. In contrast, a nonlocal event does not fizzle as distance increases. It's as if the rock's splash occurs with equal intensity in San Francisco Bay, Honolulu, Hong Kong, and all other places as well. Similarly, it doesn't matter if a person prays for someone else from far away or from the bedside. What matters is the prayer.

NONLOCAL PHENOMENA ARE IMMEDIATE

Not only are nonlocal events outside of space, they are outside of time as well. The rock's splash occurs everywhere at the same time. A mental intention does not "arrive" at a "destination," because there is no distance to cover and therefore no requirement for travel time. Moreover, as mentioned, there is no mediating energy that *needs* to travel.

These images of nonlocality offer great challenges when we first bump into them, because they defy common sense. But that is no reason to ignore them. Time and again, common sense has proved an unreliable guide about what can and will happen. Many scientists, for example, predicted that objects heavier than air could not possibly fly, that X rays were a fraud, that there would never be a market for home computers, that cellular telephones were a silly idea because people would refuse to talk in public on such a device. What our resistance to nonlocality really comes down to is that people resist the unseen. Yet as a culture we have already embraced the unseen in mind-body medicine. The mental leap has been made. Now we must go further.

CONSCIOUSNESS IS FUNDAMENTAL

The old physical, brain-based images of the mind die hard. It's as if we won't *let* the mind be nonlocal, no matter what the evidence says.

For instance, there is a tendency among specialists of the mind to try to explain consciousness in terms of something else—as some sort of subtle energy, information field, mythic field, quantum vacuum, quantum field, archetypal field. The list is virtually endless. By contrast, in Era III medicine, the mind is considered to be fundamental. *Fundamental* means that something stands on its own; it is not derived from anything else, and it cannot be explained in terms of anything more basic.

In the nineteenth century, scientists believed light needed something physical to help it along, and they invented the concept of the ether—an invisible yet undiscovered physical substance—to serve this purpose. Then they found that light was capable of traveling without assistance, and the notion of the ether was abandoned. As we shall see, evidence suggests that the mind, like light, does not need to be helped along by anything else. It is genuinely present everywhere in space and time. Since it is *already* everywhere, it has no need to "go" or be "sent" and therefore needs no sender or carrier.

Erecting symbols for the mind can be helpful as long as the symbols are not taken for the real thing. We can think about the mind *as if* it were some sort of field, and we may one day work out mathematical equations to describe such a field. The danger, however, is that we may mistake the field for the mind, confusing the map with the territory or the menu with the meal. Most brain researchers commit this mistake when they substitute *brain* for *mind.* This error is enormous, because it limits the mind not only to the brain's premises but to the brain's fate as well: destruction with physical death. The result is a deadening of humankind's spiritual aspirations, including the vision of immortality. As philosopher Michael Grosso says,

> Materialists think that consciousness is either identical with, or a by-product of, the brain. The consequences are clear. Diminish or destroy brain function, and you diminish or destroy consciousness. Verdict on immortality: death of brain implies death of consciousness; the curtain goes down forever.[14]

Grosso endorses the reality of the spirit or soul—a quality of consciousness that transcends the physical body and the ego, the sense of self we call "me." The ego resists the idea of the soul and nonlocal mind. It wants to remain unique and intact, and it struggles against being included in something larger. One might even say that the struggle over nonlocal mind is often not a debate about actual evidence but a battle fought by the ego to maintain its sense of importance.

Psychologist C. G. Jung, one of the great modern cartographers of consciousness, offered a point of view that might calm the fears of anyone who is concerned they will lose their individuality by exploring nonlocal mind. "This feeling for the infinite . . . can be attained only if we are bounded to the utmost," he wrote. "In knowing ourselves to be . . . ultimately limited—we possess also the capacity for becoming conscious of the infinite. But only then!"[15]

The evidence is overwhelming that there is *some* relationship between the brain and the *contents* of consciousness. If I take LSD or even sip a glass of chardonnay, my thoughts and feelings change as the chemicals do their work. Throughout history great thinkers have agonized about what the connection between the mind and the brain may be. Luminaries such as Plato, Schiller, William James, C. D. Broad, and Henri Bergson have proposed that the brain, rather than producing the mind, interacts with it. Grosso describes this point of view with an analogy from electronics:

> A crude analogy with radio and radio waves: the radio does not produce the radio waves; it detects, transmits, and filters them. If your

> radio breaks down, it doesn't follow that the sounds you're listening to have ceased to exist. They just cease to be detectable. An analogy is possible between this and the mind-brain relationship.[16]

René Descartes, whose influence on the Western view of consciousness has been monumental, held a similar view of the mind-brain relationship. With his famous "I think, therefore I am," he posited two distinct domains of reality, the mental and the physical. The mind and body interacted, he maintained, in the pineal gland deep within the brain. Thus for Descartes, the pineal gland functioned much like the radio in Grosso's image above.

As we explore Era III, let's bear in mind how profoundly ignorant we are about the nature of consciousness. As philosopher John Searle has said, "At our present state of the investigation of consciousness, we *don't know* how it works and we need to try all kinds of different ideas." And philosopher Jerry Fodor observes, "Nobody has the slightest idea how anything material could be conscious. Nobody even knows what it would be like to have the slightest idea about how anything material could be conscious. So much for the philosophy of consciousness."[17] If we acknowledge the limits of our current knowledge, we can remain open to the wide range of nonlocal events in our daily life.

INTERNAL SHAKE-UPS: WHY WE NEED NONLOCAL EXPERIENCES

One day when I was a university student I developed severe abdominal pain. When I couldn't ignore it any longer, I reported to the student health center. The diagnosis was acute appendicitis, and I was scheduled for emergency surgery. The operation was uneventful except for the surgeon, who was one of the most arrogant physicians I have ever met. He was utterly aloof and ignored me on his post-op rounds, confining his attention to "the incision" as if I were not present. He dressed like royalty and carried a gold-headed cane. His most remarkable eccentricity was his practice of literally "shaking up" his patients.

The first time he came to my bedside following surgery, he handed his cane to the nurse, leaned over me, placed his hands on either side of my abdomen, and without a word of explanation began to shake the bed violently. For an instant I thought he might be mad—I'd heard wild stories about the type of physician who works at student health centers. Then, leaving me to grimace in pain, he retrieved his cane and wheeled out of the room. When I could speak, I asked the nurse for an explanation for this "treatment"—which, by the way, I do not recommend. "The intestinal tract can be sluggish after an abdominal operation," she said. "Dr. ___ always shakes up his patients after surgery. He says his patients don't get adhesions if he jiggles their insides. He's trying to wake up your organs and keep things loose and flexible."

Nonlocal experiences are like that. As Grosso puts it, dreams, visions, and telepathic and clairvoyant experiences prevent "hardening of our conceptual arteries. . . . [These] barbarians from the far side of the psyche . . . inspire wonder and keep the soul young. . . ."[18] But shake-ups can be painful, and it's always tempting to lie still in our conceptual bed and not ruffle the view of reality to which we are accustomed. If we are to prevent our "mental insides" from forming adhesions and strangling reality, however, we may have to undergo a bit of painful jiggling from time to time.

THE GOLDEN RULE OF ERA III

As an example of the shake-ups that may be in store for us, consider the implications of nonlocal mind for our traditional foundations of ethical and moral behavior. Our present way of thinking is pervasive, and a good example of this is the Golden Rule, which we use as a guide for proper conduct: Do unto others as you would have them do unto you. From the ideas we have introduced so far, we can see that the Golden Rule takes a thoroughly local point of view. That is, it assumes that we are separate entities—I am I, you are you, and we are doing things to each other. But Era III recognizes the collective, unbounded nature of human consciousness. It maintains that while our bodies may be separate, our minds are not. In Era III, "I" and "other" are in some sense one, which means that what we do to others we do to ourselves. This makes possible an extension of the Golden Rule: "Do good unto others because they *are* you!"—the Golden Rule of Era III.

But growing into Era III does not mean devaluing the individual or dishonoring the private citizen. Neither does it mean renouncing our personal uniqueness and national character in favor of global sameness. Nonlocal ethics recognize that, while we are indeed individuals, we are *more* than individuals. The Golden Rule of Era III is the *fulfillment* of individuality, not its eradication.

NONLOCAL THINKING: BOTH/AND, NOT EITHER/OR

The legendary physicist Niels Bohr once said that an ordinary truth is one whose opposite is false; but a *great* truth is one whose opposite is also true. Era III, I submit, is a great truth because it reveals that we are *both* local and nonlocal, *both* individual and collective.

We see this embrace of opposites in Era III medicine. Era III medicine does not do away with the mechanical medicine of Era I or the mind-body approaches of Era II; it includes and goes beyond them. For example, surgery, a mechanical intervention of Era I, still works in Era III, but it is part of a more inclusive, comprehensive pattern. Just so, our individual, local mind is also real in Era III, but it is complemented by our universal, nonlocal consciousness. Our limitless, nonlocal mind does not destroy our individuality. The two stand side by side; both are required in describing what it is like to be conscious, to have a mind.

Some people consider this idea irreligious, yet it is implicit in religions worldwide. Consider, for example, the Christian doctrine of the Trinity, in which three individual entities—the Father, the Son, and the Holy Spirit—come together as one. The same concept is integral to certain Eastern traditions. As the modern Buddhist teacher Lama

Govinda expressed it, "Individuality is no contradiction to universality. It is an entirely wrong idea either to think of oneself as the whole world, or conversely, that one is nothing—a miserable sinner or a speck of dust."[19]

GETTING USED TO NONLOCALITY

I believe nonlocality is one of the most important discoveries humans have ever made. This concept distinguishes the science of the twentieth century from all the science that came before. I have found that even people who are unfamiliar with the term *nonlocal* often have an intuitive feel for what nonlocality is all about, and I'm convinced this is one reason why this monumental idea is beginning to infiltrate public awareness. For example, a search of the Internet turns up five hundred references for nonlocality. Still, the idea of nonlocality takes a bit of getting used to.

Alice B. Toklas, Gertrude Stein's companion of many years, relates in her autobiography how Picasso once painted a portrait of Stein in his inimitable style. Toklas told Picasso that, although she personally admired the painting, many of her friends did not. They invariably said that Stein did not look like the painting. Picasso replied, "It does not make any difference, she will."[20]

I've found that individuals who object to nonlocal mind are often like those people who didn't like Picasso's portrait of Stein: they don't think "nonlocal" looks like them. They like to think of themselves as local creatures who possess a unique personality centered inside their individual brain and body. They feel as if their mind is confined somewhere in their skull behind their eyes. The infinitude implied by nonlocal mind is too diffuse, impersonal, threatening. But as Picasso said, that does not matter; we *are* nonlocal, whether we wish to be or not.

No doubt many of those who disdained Picasso's portrait of Gertrude Stein came to like it with time. Like spinach, "nonlocal" can be an acquired taste. Similarly, I've found that if people are willing to play with the idea of nonlocality, they often become quite cordial to the idea of nonlocal mind as they realize how accurately it describes how they function.

For those of us who believe deeply in science, there are two ways of resolving questions about the nonlocal nature of our mind—to go unflinchingly through the evidence and not around it, and to permit ourselves to experience nonlocal happenings personally. This dual strategy involves the risk of transformation, which is why entering Era III and nonlocal mind is like handling intellectual and spiritual dynamite.

In the chapters that follow, I shall ask you to follow me into the scientific reasons supporting Era III medicine and a nonlocal view of the mind. But as I do so, I hope you understand that these issues involve more than data.

I once felt differently. I believed that if nonlocal mind were valid, "killer experiments" could be done that would be so persuasive they would sweep away all opposition and quell every argument. This reflected my idealized view of science—that scientists are completely rational creatures who, when faced with data, respond as objectively as a computer and "do the right thing." I no longer believe this, because I have learned that my idealized image of science was wrong. Scientists are not unemotional computers. They can be as biased and ornery as anyone else, particularly when venturing outside their field. As one respected scientist said when asked to review a scientific paper dealing with nonlocal mental events, "This is the sort of thing I would not believe, even if it were true."

In some spiritual traditions, teachers have spoken of various "eyes" that need to be opened in order for one to see certain things. In addition to one's physical eyes, which take in light, there is the eye of the mind, which is sensitive to reason, and the eye of the soul, which comprehends higher matters. In order to grasp nonlocal mind, we shall have to open all our eyes. Otherwise we may not see the evidence for nonlocal mind even when it is staring us in the face.

How can we open our various eyes to these matters? The path I endorse in this book is that of nonattachment—setting aside our ideas about how we think the world *ought* to work and letting the evidence speak to us. This sounds simple but can be difficult because of our innate tendency to put our trip on the universe.

Writing this book has given me an opportunity to relive my own journey into this area. My venture into nonlocal mind has not been a smooth, unbroken process, but a titanic struggle. I know the resistances many readers will experience—the tendency of the ego to deny a nonlocal, transpersonal dimension of the self and to protest, "These things can't happen!" This is why I have always been more at home with skeptics who demand proof than with uncritical true believers. I myself have vigorously denied the evidence for nonlocal mind for extended periods in my life. But although I am by nature and training a creature of reason, I have sometimes been reason's victim. I have used reason unreasonably—to defend my ego and my preferred version of reality instead of being open to fact.

So, in asking you to take a detached point of view in what follows, I realize I'm asking a lot. All that we treasure—our self-identity, self-esteem, and sense of individuality—rebels at the prospect of widening the dimensions of the self. But in addition to the tug of the individual self, another call has always beckoned humans—a call from the infinite, the universal, the nonlocal—and this is the call that we shall now examine.

CHAPTER 2

THE CASE FOR NONLOCALITY

The brain—is wider than the sky—
For—put them side by side—
The one the other will contain
With ease—and You —beside.

EMILY DICKINSON
Poem 632

On a cold, blustery day in December 1997, a historic meeting took place in Boston on the campus of Harvard University. Boston, cradle of the American Revolution, was hosting another revolution of sorts, a conference called "Intercessory Prayer and Distant Healing Intention: Clinical and Laboratory Research." Approximately one hundred researchers from medical schools and universities around the United States gathered to discuss experiments in these forms of nonlocal healing that they were performing at their respective institutions. All of their research studies had a common goal—to test whether or not individuals could mentally help heal distant persons who were unaware they were doing so. The researchers described this effort in various ways. Some called it "prayer," while others preferred "distant intentionality," "empathic concern," or some other more secular-sounding terminology. All the studies were ventures into Era III medicine and nonlocal mind—the capacity of human consciousness to function outside the confines of the individual brain and body.

As the proceedings flowed from morning into night, I could hardly believe this event was happening. A few years earlier it would have meant professional suicide for a researcher to perform experiments that took seriously the possibility that healing could be freed from the individual mind. But here were world-class scholars, highly respected in their fields, discussing their latest findings as forthrightly as if nonlocal mind were a new treatment for AIDS (which, as it turned out, it is, as we'll see). What a contrast to my first ventures into this area! When I first began my explorations of nonlocal mind years ago, I did so with a sense of dread. In my medical training I had been warned that this field was "pathological," and that flirting with "ESP" would damage a medical career. Researching the connections between health and religious devotion would unleash what a colleague of mine called the ATF—the anti-tenure factor.[1] But my curiosity had got the best of me. Like a teenager sneaking into an X-rated movie, I had disregarded the warnings and had avidly begun to explore this field, but always with trepidation. Now it looked as if things were finally loosening up.

It was fitting that the Harvard meeting was being held in the Swedenborg Chapel, an elegant Gothic structure named for Emanuel Swedenborg (1688–1772), the Swedish scientist and philosopher who was one of the most respected scholars of his era. Superbly educated in science and mathematics, he lectured in London and at Oxford. He specialized in mining and metallurgy and was royal engineer to the king of Sweden. He founded Sweden's first scientific society, recommended decimal coinage, and made original contributions in a variety of fields including algebra, geology, and crystallography.[2]

Swedenborg was no stranger to nonlocal mind. In July 1759, while a guest at a large dinner in the home a wealthy merchant in Gothenburg, Swedenborg suddenly became quite agitated at about six in the evening. He announced that in Stockholm, three hundred miles away, where he lived, a great fire had broken out. He described its location and when it began. He regained his composure only after "seeing" that the fire had stopped before reaching his house. He reported the fire to the governor of Gothenburg the next day. News of the fire reached the city by way of normal channels the following day, confirming the details of Swedenborg's report.[3] Events such as these were commonplace in Swedenborg's life. He would have been at home at the Harvard meeting, and his spirit hovered like a blessing over the proceedings.

Another historic connection between nonlocal mind and the Harvard conference was through Edgar Mitchell, the Apollo 14 astronaut who had walked on the moon. After returning from the mission, Mitchell founded the Institute of Noetic Sciences, with which Dr. Marilyn Schlitz, who convened the Harvard meeting, is affiliated as director of research. Mitchell's purpose was to explore the nature of consciousness, which was a passion of his. While in space he performed an experiment on the sly to test nonlocal mind, which he describes in his book *The Way of the Explorer.* His comrades did not know what he was up to, and only four people on Earth were aware. "Every evening as the crew settled in for an attempt to sleep in zero gravity and the cabin grew quiet, I would take a moment and pull out my clipboard on which I had copied a table of random numbers along with the five Zener symbols made popular by [parapsychology researcher] Dr. J. B. Rhine: a square, circle, star, cross, and wavy line. . . . Meanwhile, through tens of thousands of miles of empty space, my collaborators in Florida would attempt to jot down the symbols in the same sequence that I had arranged on the clipboard."[4] The results were significantly different from what would be expected by chance, suggesting that information can be shared nonlocally between distant individuals, without any sort of sensory exchange.

Many of the studies the scientists presented at Harvard have since been published in professional journals. We shall review some of them in the pages that follow.

The Harvard meeting began with a bow to history. The opening session featured two pioneers in distant healing research, psychologists Lawrence LeShan and Bernard Grad. It was a touching gesture, honoring the elders who helped establish this field.

LeShan is a legend in the field of Era II or mind-body medicine. He was one of the first researchers to search for links between cancer and the mind. These connections were considered heretical when he began to explore them in the years following World War II, but LeShan has seen his intuitions vindicated. Today it is virtually taken for granted that there are important linkages between the mind and the neurological and immune systems. A new medical field now exists, psychoneuroimmunology, which rests on LeShan's instincts and the premises he laid down. As Jeanne Achterberg, the internationally recognized researcher in mind-body medicine says, "All trails in this field lead back to LeShan."

LeShan was also one of the earliest explorers of nonlocal healing, a field he entered in the 1960s. His seminal book on distant healing, *The Medium, the Mystic, and the Physicist,* first published in 1966, influenced me greatly.[5] When my wife, Barbara, and I were married in 1972, I took LeShan's book along on our honeymoon. I knew nothing about this field but was intrigued by the book's title, and I thought a little diversionary reading might be a good idea. LeShan's descriptions were spellbinding. Honeymoons can be memorable for many reasons, but LeShan, I am sure, helped make mine more memorable than most.

EVIDENCE OF NONLOCAL HEALING

The other patriarch of distant healing featured in the meeting's opening session was psychologist-researcher Bernard Grad, of Montreal's McGill University. Grad's research, which he first began to publish in the early 1960s, placed nonlocal healing on a scientific footing. I regard his experiments as of Nobel quality and significance, because they opened an important field to experimentation and practical application. They are ingenious and profound and have set a standard for all the experiments in nonlocal healing that have followed.

Great science does not become stale. Although Sir Isaac Newton's and Galileo's discoveries are more than three hundred years old, they are still used to launch satellites into orbit. Similarly, Grad's experiments, although performed thirty years ago, remain as relevant and up-to-date as if done yesterday. All current studies rest on the principles they uncovered. So, before examining some of the healing experiments that came out of the Harvard conference, let's look briefly at Grad's breakthroughs to set the stage for them.

Biologist Lyall Watson once proposed that the inanimate world sometimes comes alive, rather like those cartoons in which the toys in the toymaker's shop spring to life when he shuts things down for the night. In some strange way they take on our "emotional fingerprints," as Watson put it, catching our moods and emotions.[6]

Grad was one of the first researchers to test seriously the hypothesis that our moods might affect the things around us, along the lines Watson suggested. Grad tested whether or not mental depression might produce a negative effect on the growth of plants.[7] This idea fits with the common belief that some people have green thumbs and that one's thoughts and emotional states may play a role in how vigorously plants grow. Grad theorized that if plants were watered with water that had been held by depressed people, they would grow more slowly than if watered with water held by people in an upbeat mood. In a controlled experiment he tested three subjects: a man known to have a green thumb and two patients in a psychiatric hospital, one a woman with a depressive neurosis and the other a man with a psychotic depression. Each person held a sealed bottle between his or her hands for thirty minutes, and the water was then applied to barley seeds. The green-thumb man was in a confident, positive mood at the time he held his solution, and his seeds grew faster than those of the others and the controls. Unexpectedly, the normally depressed, neurotic woman responded to the experiment with a brighter mood, asking relevant questions and showing great interest. She cradled her bottle of water in her lap as a mother might hold a child. Her seeds also grew faster than those of the controls. The man with the psychotic depression was agitated and depressed at the time he held his solution; his seeds grew slower than the controls. Grad's study suggests that (1) emotions may influence healing both positively and negatively, and that (2) physical objects may mediate these influences, perhaps by picking up on the moods of humans, as Watson suggested, through a mechanism not yet explained by science.

Grad performed a similar experiment with a healer who used his thoughts and prayers to make things well. Barley seeds were damaged by watering them with a 1 percent saline solution. The healer, in an attempt to mitigate the retardant effect of the saline, treated the beaker of saline with which the seeds in the experimental group were watered by holding it in his hands. He did not treat the saline used to water the control seeds. The experiment was repeated three times with similar results: seeds watered with the treated saline germinated and grew better than those watered with untreated saline. The odds against chance as an explanation for the results of the three experiments were one in twenty, one in fifty, and one in a thousand. These studies suggest that healing intention may somehow be mediated through a secondary vehicle or agent, as in the green-thumb studies above.

When I first read about these experiments, I thought about the physical objects doctors and nurses use as go-betweens in their interactions with patients—stethoscopes, bandages, monitors, syringes, and needles. Could these objects take on our moods and affect how our patients respond? Some nursing tasks are a dead ringer for Grad's experiments. For example, nurses hold IV solutions in their hands before "watering" their patients with them. Are the patients affected by the mood of the nurse via the IV she or he administers? Can a stethoscope transmit a doctor's foul mood to a patient if the doctor is having a bad day? And if negative moods can be communicated, why not positive ones? I once knew an elderly hospital janitor who was regarded by the nurses as something of a shaman. The man repaired watches in his spare time; he simply adored gadgets. Instruments that had broken down would often start working again in his presence, without his touching them. Once, when I was attending a sick patient in the middle of the night, a cardiac monitor malfunctioned. "Call Robert," the nurse in charge instructed. "Ask him to stand in the room." He did, and the monitor resumed function.

In another experiment, Grad studied mice in whom goiters had been produced by withholding iodine from their diet and by giving them a special drug. Seventy mice were divided into three groups: (1) a baseline-control group, which received no treatment for their goiters; (2) a group of healer-treated mice, which were held in special boxes by a healer for fifteen minutes twice daily, five days a week, and once on Saturdays; and (3) a heat-control group, kept in cages heated to the same temperature and for the same length of time as the "held" group. The rate of thyroid growth was measured by weighing the thyroids of the mice after they had been sacrificed. The thyroids of the healer-treated group of mice grew significantly more slowly than those of both the controls. The odds against a chance explanation for the experiment's outcome were less than one in a thousand.[8]

In a variation of this experiment, Grad divided thirty-seven mice on a goiter-producing diet into treatment and control groups. This time the healing was again given via an intermediary object, fleecy cotton wool, held in the hands of the healer for fifteen minutes once the first day and twice for the next twenty-four days of the study. Ten grams of cotton wool were placed in each cage with four or five mice for one hour, morning and evening, six days a week. Mice in contact with the healer-held intermediate substance developed goiters significantly more slowly than the controls. Again, there was less than one chance in a thousand that these results were due to chance.[9]

Grad was the first person to study the effects of nonlocal healing intention on the healing of wounds. He anesthetized forty-eight mice and created uniform surgical incisions on their backs by removing a piece of skin about one-half by one and a half inches. A healer held the cages of one-third of the mice for fifteen minutes twice daily while trying to heal them by mental means. One-third of the mice were placed in cages that were adjusted to the same temperature as the cages being held by the healer. The remaining one-third of the mice served as controls, being moved about as the other two groups but receiving no healing intent and no additional heating. The rates of healing of the wounds of all the mice were assessed by tracing the wound shape on paper, cutting out the tracing, and weighing the cutout. After fourteen days, the treated group had healed with significantly greater rapidity than the control group, with less than one chance in a thousand that the results were due to chance.[10]

Following these wound-healing experiments, a more elaborate study was undertaken by Grad in coordination with Dr. R. J. Cadoret and Dr. G. I. Paul of the University of Manitoba. A total of three hundred mice in three treatment groups were used: one hundred mice treated by the healer, one hundred mice treated by medical students who claimed no special healing abilities, and one hundred mice given no treatment. The study was double-blind, meaning that no one knew which group received the healing influence and which did not. At two weeks following wounding, the area of the wounds were significantly smaller in the mice treated by the healer, compared to the control group and the group treated by unskilled medical students.[11]

Grad's studies don't tell us *how* these events happen, but they rebut the contention that distant healing effects are always due to the placebo effect. Placebo effects are improvements due to what a subject expects or thinks will happen—the results of suggestion, the power of positive thinking. Seeds, plants, and mice presumably don't think positively or negatively and are not susceptible to the effects of suggestion or expectation. This means that the placebo effect does not apply to them. The fact that Grad showed these events taking place in animals as high on the evolutionary scale as mice and as low on the scale as seeds points to the fundamental nature of whatever it is that is producing the effect. Thus Grad concludes, "[These] phenomena . . . throw new light on the basic unity of man, animal, and plant. . . ."[12]

Grad helped change the popular perception that nonlocal mind is only about card guessing and mind reading. He showed that our thoughts and intentions affect living things nonlocally in medically significant ways, such as the healing of wounds and the growth of tumors. That is why Grad is one of the architects of Era III medicine and why he was honored at Harvard.

PRAYING FOR PATIENTS WITH AIDS

The studies discussed at the Harvard conference followed in Grad's tradition. One of them dealt with one of the greatest medical challenges of the modern era: AIDS.

Dr. Elisabeth Targ and her colleagues at California Pacific Medical Center in San Francisco tested whether distant healing (DH), including prayer, has a therapeutic effect on health in AIDS when subjects do not know they are receiving treatment. The four-member team brought impressive credentials to the task. Targ, a physician-psychiatrist, is the director of Psychosocial Oncology Research and is on the clinical staff of the University of California, San Francisco, School of Medicine; Dr. Helene S. Smith, who has since died, was director of the Geraldine Brush Cancer Research Institute and an associate professor in the Department of Medicine at the UC San Francisco School of Medicine; Fred Sicher is director of the Sausalito Consciousness Research Laboratory; and Dr. Dan Moore II is associate professor in the Department of Statistics, UC San Francisco School of Medicine.

I had often talked with Targ about the best way to evaluate nonlocal healing, and I had followed the progress of her study since its inception. When she told me that she and her colleagues had chosen AIDS as the disease they wished to test, I groaned.

"Why pick a disease for which there is no known cure?" I asked Targ. "This isn't being fair to the healers. Why not pick the flu?"

"Why are you so cowardly? I thought you believed in this stuff!" she chided. "AIDS is perfect. This way, the skeptics can't say we picked a disease that was too easy. Besides, healers like a challenge!"

Targ was right, as her experiment showed.

The controlled, randomized clinical trial used the same rigorous scientific standards that are required for testing a new drug. Forty patients with advanced AIDS—thirty-seven men and three women with a mean age of forty-three—were recruited from the San Francisco Bay Area through advertisements and fliers. They were from various ethnic and cultural groups. All the patients received standard medical care for their illness. Twenty, however, received distant healing intentions in addition. The study was also double-blind, so no one involved, including the patients and the scientists, knew who was in the distant healing group.

Forty volunteers throughout the United States and Canada conducted the distant healing therapy. Each healer was given a patient's first name and photograph to help that healer develop a personal connection with the subject. Healers were asked to focus their mental energies on the patient's health and well-being for an hour a day, six days a week, for ten weeks. Healers were from eight different healing traditions, including Christian, Jewish, Buddhist, Native American, and shamanic practices, as well as graduates of bioenergetic and meditative healing schools. The healers had an average of seventeen years' experience. They were organized on a rotating schedule so that each patient was treated by a different healer each week.

The patients were assessed by blood tests and psychological testing at the beginning of the study and at the end of the six-month follow-up period. No differences were seen in the $CD4^+$ (a type of immune cell important in resisting the AIDS virus) counts between the distant healing and control groups. But a blind review of their medical charts revealed several significant differences. Those patients who had received distant healing intentions had undergone significantly fewer new AIDS-related illnesses, had less severe illnesses, required fewer doctor visits, fewer hospitalizations, and fewer days of hospitalization. Moreover, those receiving distant healing showed significantly improved mood compared with controls. The psychological tests showed that the treatment effects *were not affected by the subjects' beliefs about which group they were in.* This study confirmed the results of a prior pilot study involving half as many patients, which had produced similar findings.[13]

PRAYING FOR BYPASSES IN BOSTON

One of the most ambitious studies discussed at the Harvard meeting is an ongoing experiment headed by Dr. Herbert Benson, of the Mind/Body Medical Institute at Boston's New England Deaconess Hospital. Benson has spent years studying the *local* effects of prayer and meditation—how a person's own body responds when they relax. Now he is investigating prayer's *nonlocal* effects on distant individuals.

The experiment involves twelve hundred patients undergoing coronary artery bypass surgery in three different Boston hospitals. One-third of the patients are to receive prayer from Carmelite nuns and other religious groups, without knowing for certain whether they are being prayed for; one-third are to receive prayer and will be informed that they are the subjects of prayer; and one-third of the patients are not to be prayed for. The first two groups hopefully will show whether prayer has an effect and whether such an effect is heightened if one is actually aware of the prayers of others.

The fact that nonlocal healing effects were being studied at Harvard Medical School spoke volumes. As one researcher commented, "If you can study prayer in Boston, you can study it anywhere." This is a telling point. Critics have claimed that studies in distant healing are done at fringe institutions by inept investigators. The fact that these studies are being conducted at premier research institutions by respected, well-known scientists has eroded the psychological and scientific resistance to this field.

"CARDIO-SPIRITUAL" HEALING AT DUKE

Shortly after cardiologist Mitchell Krucoff and nurse-practitioner Suzanne Crater presented their study at the Harvard meeting, it became the focus of national attention in *Time* magazine.[14] Krucoff and Crater are impressively credentialed. Krucoff is an associate professor of medicine and cardiology at Duke University Medical Center and is director of the cardiovascular laboratories at the Durham VA Hospital, one of Duke's main teaching hospitals. Crater's field is also medicine and cardiology. Together they coordinate the MANTRA project, a "cardio-spiritual" program combining high-tech cardiology and intercessory prayer, music, mental imagery, and touch. MANTRA is an acronym for "Monitoring and Actualization of Noetic Training." These interventions were studied on patients undergoing angioplasty, a coronary artery dilating procedure. Preliminary results with thirty patients were presented to cardiologists at the American Heart Association's annual 1998 meeting. Although the number of patients is too few to draw statistically firm conclusions, the outcomes of the prayed-for group were 50 percent to 100 percent better than those of a control group not receiving prayer. The study is being expanded to include fifteen hundred patients at five different medical centers in San Diego, Washington, D.C., and Oklahoma City, in addition to Durham.

I like Krucoff and Crater's study for a special reason. Critics sometimes say that researchers who investigate prayer have a hidden agenda—trying to advance their private religious beliefs. The MANTRA project eliminated this objection by recruiting intercessory prayer from a variety of religious groups. When people are assigned to the prayed-for group, their names are given to the Carmelite Sisters of Baltimore, Maryland, and are e-mailed or phoned to Buddhist monks in Kopan Monastery, Nepal, and to Nalanda Monastery, France. The patient's name is also entered on the Virtual Jerusalem Web site, whose staff actually inserts written prayers for them in cracks in the Western Wall of that city, according to ancient Jewish custom. Then the patient's name is forwarded to Silent Unity, an interdenominational Christian prayer group in Missouri, and to Baptist, Moravian, and Church of Abundant Life congregations in North Carolina. All these groups pray for the patients collectively, at the frequency and duration that is customary for them. This way, no single religion can claim credit, and the experimenters cannot be accused of fronting for any particular religion.[15]

PRAYER AND THE PRAY-ER

I discovered early in my investigation of prayer research that members of the clergy almost never perform scientific studies to evaluate the effects of prayer. Their laboratory is their own life and the lives of those they serve, and they believe that empirical, scientific proof is unnecessary. Some actually see science as the enemy of faith. Others go even further and say it is a sin to "test God."

An exception is Father Seán O'Laoire, a Catholic priest and psychologist who holds an undergraduate degree in mathematics and a Ph.D. in transpersonal psychology. Father Seán works in the Catholic community of the San Francisco Bay Area, where he has a private counseling practice. His research on intercessory prayer, published in 1997, was presented at the Harvard conference.[16]

O'Laoire's study divided 406 individuals into two groups, one of which received prayer and one that did not. He also examined the effects of prayer on the ninety agents who did the praying. His experiment, like those above, was a controlled double-blind study in which no one knew who was and was not receiving prayer. The goal was to examine the effects of prayer on eleven measures of self-esteem, anxiety, and depression. O'Laoire found that the subjects being prayed for improved on all eleven measures. And here is the unique feature of O'Laoire's study: on ten of the eleven criteria, O'Laoire discovered that the agents doing the praying improved more than the subjects for whom they were praying.

This does not come as a surprise to many people who pray. In discussing prayer with groups of laypeople, I often ask them why they pray. Invariably someone says, "Because it makes *me* feel better!" O'Laoire's study shows they are not fooling themselves.

Surgeons don't report that removing their patient's appendix results in *their* improvement. Yet, O'Laoire showed that Era III therapy is good for the therapist as well as the patient—one of the most unique features of Era III medicine.

DISTANT INTENTION AT MT. SINAI SCHOOL OF MEDICINE

One of the most ancient ways in which people have tried to heal others at a distance is through the technique of qigong (pronounced "chee-gong"), which has more than a three-thousand-year history in China. This healing method is centered around the concept of a vital force, or qi, which is believed to permeate all living systems and indeed the entire world. Good health, it is claimed, involves an unobstructed and balanced flow of qi.[17]

With the recent increase in cultural exchange between China and the West, qigong masters have begun to filter into the United States. Two qigong masters, Ronger Shen and Yi Wu, were among the first in China to learn a technique called Soaring Crane qigong, which was introduced to the public in China in the 1980s and which soon attracted over twenty million adherents. When Shen and Wu came to the United States they collaborated with a research team at Mt. Sinai School of Medicine to test the effects of qigong in a sophisticated, well-controlled experiment published in 1994.[18]

In addition to the qigong practitioners, the Mt. Sinai team was composed of David J. Muehsam, M. S. Markov, Patricia A. Muehsam, and Arthur A. Pilla. As their "subject" they chose a biochemical reaction involved in the contracting of muscles that line the blood vessels and intestinal tract. The biochemical reaction is highly complex and occurs in stages. It requires the binding of calcium to a protein called calmodulin, which then activates an enzyme called myosin light chain kinase. This activated enzyme then causes phosphorous to attach to molecules called myosin light chains. The end result is the production of energy required for the contraction of muscle tissue.

The researchers asked the qigong practitioners to treat the tissue samples, which were taken from animals, as they would treat a patient in real life. They stood two to five feet away from the test tubes during treatment, which lasted for six minutes for each sample. In all of nine trials, the qigong masters were able to modify the biochemical reaction by an average of 15 percent, which is an effect size seen in many clinically significant biological reactions in the body. The odds against a chance explanation of the outcome were less than one in twenty.

In another phase of the experiment, one sample was treated by an untrained individual, resulting in the smallest effect seen during the experiment.

On balance, these findings suggest that the distant effects of mental intention are real, that mental intention can significantly affect biologically important biochemical processes, and that no physical contact between the practitioner and the sample is required.

These results do not stand alone. Dr. Garret L. Yount, a cancer cell biologist at the Geraldine Brush Cancer Research Institute at California Pacific Medical Center in San Francisco, has also worked with a qigong master from Asia. Yount's "subject" is cervical cancer cells, which the qigong master is asked to interact with at a distance in an attempt to neutralize or kill them. Preliminary results indicate that the qigong practitioner can do so.[19]

The Mt. Sinai experiment and that of Yount reveal a side of nonlocal mind that often goes unnoticed. When we think of the power of consciousness in health, we often equate it with making something stronger or healthier. Yet these two studies show that mental intentions, in addition to increasing the vitality of living things, can also inhibit and destroy. Many people are horrified to contemplate the possibility that the mind can kill. But we frequently encounter situations where this is precisely what we need our minds to do. For instance, when we pray for an individual to recover from AIDS or pneumonia, we are asking that the viruses and bacteria causing the infections be put out of commission. When we pray for someone to get well from cancer, we want the cancer cells to die. If nonlocal mind *can't* harm, it will not be as effective in healing as we desire it to be.

When I first began to explore the field of distant intentionality, I bumped into scores of studies such as those above that were performed on cells, microbes, animals, or biochemical reactions instead of human beings. At first I was puzzled about why researchers would not use people, since they are the objects of healing in real life. But I quickly saw that these strategies make sense. If the experiments work in nonhumans, the results cannot be dismissed as due to the placebo response, the power of positive thinking, because microbes, plants, and animals presumably do not think positively *or* negatively. Moreover, studies in nonhumans can be done with greater precision than in humans, because fewer variables are involved. In any case, the fact that animals and microbes are used as test subjects is really not surprising, considering how medical research is routinely done. Medical scientists, when researching a new drug such as an antibiotic, first test it on bacteria, animals, and human cells growing in tissue cultures. When they use nonhumans as their subjects, researchers in distant intentionality are merely following the same logic.

VISUALIZATION PROTECTS RED BLOOD CELLS

A 1990 study by William G. Braud at the Mind Science Foundation in San Antonio, Texas, is another example of using human tissue in testing the effects of distant intentions. Braud tested whether thirty-two unskilled individuals could mentally protect, at a distance, human red blood cells from hemolysis—swelling and bursting—when they were placed in a weak salt solution. Each person tried to do so using visualization techniques for ten test tubes, while ten additional tubes served as the uninfluenced controls. As an aid to visualization and intention, the subjects were shown a color slide of healthy, intact red blood cells. The rate of hemolysis was measured photometrically, which involves passing a light beam through the solution. The subjects and the tubes were placed in different rooms, and the experimenters and technicians performing the measurements were blinded as to the identity of the treated and control tubes. Results indicated that the distant intentions were effective: The red blood cells in the treatment test tubes did not dissolve nearly as much as the controls. The possibility that these results could have occurred by chance alone was less than one in five thousand.[20]

Braud's experiment resembles the first step in testing a new drug: see if it works in the test tube. His findings may now be ready to apply to actual medical problems. There are several diseases called hemolytic anemias in which an individual's red blood cells fragment and dissolve too readily. A logical next step would be to test whether this process could be controlled through healing intent, by either the patient or someone else.

PRAYER IN THE CORONARY CARE UNIT

The best-known study in nonlocal healing is that of cardiologist Randolph Byrd, which took place at San Francisco General Hospital.[21] Byrd's 1988 study is the most celebrated experiment in distant healing in the twentieth century, and its impact has been enormous. It extended Grad's pioneering work to the human level, and it focused on our biggest killer: heart disease.

Byrd, a devout Christian, believes his experiment was divinely inspired. "After much prayer," he states, "the idea of what to do came to me."[22] One day he was walking down a corridor of San Francisco General Hospital, discussing a cancer case with another physician. His colleague, despairing, said, "I wish there were something more we could do!" Byrd states, "From my past experience, I should have known better than to say what I did: 'We could try prayer.'" His colleague stared at him and then resumed walking, now rapidly. Finally he responded, "I meant, Dr. Byrd, something *scientific.*"

Byrd realized that he, like many physicians, was being confronted by two of his deepest commitments: to science and to his faith. He believed deeply in both. Could they be joined? Could one do a scientific study on prayer? What if the study didn't work?

"Doctor," he replied, as he and his colleague parted in the hospital's halls, "you've given me an idea." Driving home that night along California's Highway 101, the idea began to take shape. Four weeks later his proposed experiment had been approved by the hospital's research committee. But for practical reasons he needed the approval from one of the hospital's leading cardiologists in the coronary care unit.

"Let's see if I have this right, Dr. Byrd," the cardiologist said. "You want to run a study on the therapeutic value of *prayer?"* Byrd nodded. "You'd publish your results in a medical journal?" Byrd nodded again. "What if you pray and it makes no difference? Would you print that, too?"

"Yes," Byrd said. "I will publish in either case."

"Go ahead then," he agreed. "And good luck."

Ten months later Byrd had made medical history.

Byrd and Janet Greene, his research assistant, asked each patient in the six-bed coronary care unit whose condition was stable if he or she wanted to help study the effect of prayer on treatment. "Answers ranged from an enthusiastic 'I'd really like that!' to an 'I guess it can't hurt' to one indignant refusal," Byrd relates.[23] Over a ten-month period, a computer assigned 393 patients who agreed to participate to either a group that was prayed for by home prayer groups (192 patients) or to a group that was not remembered in prayer (201 patients). The study employed safeguards that are characteristic of good clinical experiments, including randomization and double-blind precautions in which neither the patients, nurses, nor doctors knew which group was which. Byrd recruited members of several Protestant and Roman Catholic groups from around the country to pray. They were given the first names of their patients as well as a brief description of their diagnosis and condition. They were asked to pray each day but were otherwise given no instructions about how to do so. "Each person prayed for many different patients," Byrd explained, and "each patient in the experiment had between five and seven people praying for him or her."

The patients receiving outside prayer differed from those receiving no prayer in several ways:

1. They were five times less likely to require antibiotics (three patients compared to sixteen patients).
2. They were three times less likely to develop pulmonary edema, a condition in which the lungs fill with fluid as a consequence of the failure of the heart to pump properly (six compared to eighteen patients).
3. None of the prayed-for group required endotracheal intubation, in which an artificial airway is inserted in the throat and attached to a mechanical ventilator, while twelve in the group not prayed for required mechanical ventilatory support.
4. Fewer patients in the prayed-for group died (thirteen compared to seventeen patients, a difference that was not statistically significant).

Because Byrd's study has been so influential, a digression is in order to examine it in greater detail.

Byrd's results were so striking that even a few skeptics were intrigued, such as the late Dr. William Nolen, a surgeon who had written a book debunking faith healing. "It sounds like this study will stand up to scrutiny," he said. Nolen, a practicing Catholic, did not pray for his patients but said that perhaps he should, based on Byrd's findings. "If this is a valid study," he said, "for God's sake, maybe we doctors ought to be writing on our order sheets, 'Pray three times a day.' If it works, it works."[24]

Byrd's inner conflict between science and religion mirrors the tension felt by many if not most physicians. Most professionals try to ignore or compartmentalize these twin tugs. This leads to a schizophrenic existence—one part of life dedicated to materialistic science, another part to the transcendent or divine. Byrd did not believe this split was necessary. His results certainly bear out his beliefs.

When I first learned of Byrd's study, I was deeply involved in mind-body issues, the medicine of Era II. I had little interest in prayer, and it never really occurred to me to pray for my patients. But after reading Byrd's study I could not put it out of my mind. What if he were right? If I did not pray for my patients, was I withholding something valuable? Was I justified in *not* praying? I pondered these questions for months, becoming increasingly uncomfortable. Eventually I decided to resolve this issue the best way I knew how—by searching out all the scientific studies that had ever been done on prayer and distant healing. I assumed this would be quick work, for I had never heard of such studies and assumed there were only a handful of them. I did not realize that I was beginning a search that would require several years and that I would uncover nearly 150 experiments beyond Byrd's. As my search proceeded, a new world opened. Soon I was convinced that distant intention and prayer were among the best-kept secrets in medicine. Not only did I begin to pray for my own patients, I began to examine additional ways in which consciousness manifests nonlocally in our lives. So I owe a debt to Byrd I have never acknowledged. His study changed my life and my practice as a physician, and I am grateful.

CRITICS OF INTERCESSORY PRAYER

Byrd's study aroused a furor among critics of prayer. In general, these responses come from people who believe that prayer contradicts a materialistic view of the world, to which they are deeply committed. They sometimes imply that prayer researchers have a hidden agenda, trying to promote their personal religious beliefs under the guise of science.

Critics often complain that it is impossible to accurately study prayer in patients who are seriously ill. When people are facing death, as many of Byrd's heart attack patients were, they pray for themselves, and their loved ones pray for them. This means that it is impossible to establish a pure control group, which, by definition, is not supposed to receive the treatment being evaluated. Byrd's study, they say, did not test prayer versus no prayer; instead, it tested differing degrees of prayer. To complicate matters, prayer cannot be quantified; there is no way to measure a "dose" of prayer. This would be like testing a drug when nobody knew how much the patients were taking. If one can't determine how much prayer the subjects are receiving, there is no way to draw conclusions.

Moreover, critics emphasize the small differences favoring the prayed-for group. Although fewer patients died in the group receiving assigned prayer (thirteen compared to seventeen in the control group), the difference in mortality was not statistically significant, as mentioned. Even in areas where statistically significant differences were found, the superiority of the prayer was not overwhelming: only 7 percent fewer patients required antibiotics; 5 percent fewer required diuretics; and there were 5 percent fewer cases of pneumonia, 6 percent fewer instances of congestive heart failure, and 5 percent fewer instances of cardiopulmonary arrest. Overall, 85 percent of the prayer group were judged to have a "good hospital course" after entry, compared to 73 percent of the control group. The most impressive difference was the need for intubation and mechanical breathing support: none of the prayed-for group required such care, while twelve of the controls received this intervention. Thus, except for one category of clinical response, the prayed-for patients achieved only a 5 to 7 percent improvement over the controls.

A word of caution: We should be careful not to adopt a double standard in which we demand more from prayer than from penicillin or Prozac. Physicians usually don't know why any drug is not more powerful than it actually is or why it works for one person and not another. When a drug does work, we are grateful; we don't waste time condemning it because it isn't 100 percent effective. Indeed, no therapy known to humankind works all the time. It is possible to look at a glass as half empty or half full. The truly remarkable point may be that prayer works *some* of the time, not that it doesn't work *all* the time.

Although I was initially critical of the "mere" 5 to 7 percent advantage demonstrated by the prayed-for patients in many areas of Byrd's study, I believe this deserves another look. Compared to studies in other areas of medicine, this level of improvement is sensational. Some discoveries in medicine that were minuscule in comparison have been heralded as breakthroughs. Consider, for example, the results of a meta-analysis of twenty-five medical studies investigating whether aspirin helps prevent heart attacks, published in the prestigious journal *Science* in 1990.[25] (A meta-analysis is a mathematical tool used by researchers to discover trends and patterns in large numbers of studies that may not be obvious when the experiments are analyzed individually.) When the "confidence levels" of the individual studies were examined, only five of the twenty-five studies showed aspirin to be of value, while 80 percent were failures. Anyone skeptical of aspirin's ability to prevent heart attacks might go away unimpressed. Yet when the results of all the studies were combined, the aspirin effect was found to be real, and it is heralded today as a major advance in preventing heart attack. These results are statistically less significant than Byrd's, yet because they don't address an issue as sensitive as prayer, we have no resistance to embracing and heralding the results. The magnitude of the prayer effect in Byrd's study is gigantic compared to the aspirin effect. Something other than science is at work in criticizing the Byrd study.

As to the concern that researchers can't establish a pure control group when studying prayer in sick humans, valid conclusions can still be drawn in such situations. There are many examples in conventional medical research in which the control group receives the same medication as the treatment group, although in differing amounts, as when one tests high-dose versus low-dose therapies of various sorts. I have discussed the problem of unassigned, outside prayer for the control group with several epidemiologists who are experts in experimental design. They suggest that "the problem of extraneous prayer" should balance out between the treatment and the control groups when large numbers of patients are involved. The treatment group, they say, should be flooded by more prayer by virtue of the prayer groups assigned to pray for them. But how can researchers calibrate the *dose* of prayer—its quality, intensity, and fervor? The experts acknowledge this problem but assert that this factor should also balance out between the treatment and control groups in studies involving large numbers of patients.

It is possible to overcome all the major resistances to prayer research by doing the experiments in nonhumans—for example, praying for bacteria in test tubes to grow faster than unprayed-for controls. In these experiments, one can safely assume one has a pure control group—that is, the control bacteria are not praying for themselves, and their fellow bacteria are not contaminating the control group with extraneous prayer. Or when surgical wounds heal faster in mice that are prayed over, as in Grad's experiment above, the improvement is presumably not due to their own positive thinking.

The idea of extending prayers, thoughts, and intentions to nonhumans seems quite natural to many people. Following the 1993 publication of *Healing Words,* my initial book about prayer and healing, I received an outpouring of letters from veterinarians. In story after story they told me how they prayed for their patients as if they were human. I also received letters from microbiologists who described similar attitudes toward microbes, and from botanists who described intimate attitudes toward plants. These letters fit with my experiences growing up on a farm in central Texas. All the farmers I knew, including my parents, were on nonlocal terms with plants and animals. They were incessantly praying for a sick animal to recover or were trying to nudge crops along with their thoughts and intentions. So it seems natural to me that plants and animals could substitute for humans in healing experiments.

Ironically, healing experiments in nonhumans have illuminated differences among the healers. This is an important aspect of the evidence favoring nonlocal mind. If distant intentions do not work, then there should be no significant differences in the success rates of healers and pray-ers in actual experiments. *All* the healers should fail; none should consistently stand out over any other. Yet that is not what we see. Some healers rise to the top when put to the test, while others sink.

Moreover, the healers who excel are the ones we would expect to do so—those who believe in the power of healing intentions and who have cultivated this practice in their personal life, as in the following experiment.

STIMULATING THE GROWTH OF YEAST CELLS

In 1995 Dr. Erlendur Haraldsson, professor of psychology at the University of Iceland, described an experiment he performed with Dr. Thorstein Thorsteinsson, a biochemist on the faculty of medicine at the same institution.[26] Seven subjects took part in the study—two spiritual healers, one physician who believed in prayer and who used it in his practice, and four students with no experience or particular interest in healing. All seven subjects tried to stimulate the growth of yeast cells in ten test tubes, with ten as controls. They were not allowed to touch the test tubes or to come closer than one foot. Then all the tubes were stored in the same place for twenty-four hours, after which the growth of the yeast was measured in each tube by sophisticated methods commonly used by microbiologists. The experiment was well designed and properly controlled, with the technicians blinded as to which tubes were which. Overall, the treated test tubes showed greater growth than the controls. The bulk of the positive influence, however, was due to the three subjects who were actively involved in healing in their lives. The students, who professed neither skills nor interest in healing, scored at chance levels. When the scores of the healers were separated from those of the students, the odds were less than two in ten thousand that they could have occurred by chance.

The University of Iceland experiment makes sense. Experience, skill, and interest count in any human endeavor, including experiments in distant healing.

Are there born healers, just like born surgeons? Perhaps one day we shall devise ways of identifying them. Will we find ways of credentialing them also? Will we eventually recognize board-certified healers, just as we honor board-certified neurosurgeons and cardiologists?

NONLOCAL MIND IN ANIMALS

We've all heard the fantastic stories of lost animals reuniting with their owners. There is the case of Minosch, a German cat, who reportedly traveled fifteen hundred miles in sixty-one days to return home after being separated from his vacationing family. Or there was Bobby, a pedigreed collie with a quarter strain of Scotch sheepdog, who got lost on a family trip in Indiana, but found his way to the family's new home in Oregon, where he had never been, three thousand miles away, crossing the Rocky Mountains and several ice-choked rivers in the dead of winter to do so.[27]

Thousands of similar cases have been reported. No doubt some can be dismissed as involving look-alike animals, but not all. Often the returning animal has its original collar in place, as in Bobby's case, and can be further identified by distinguishing marks and scars. I once cited Bobby the collie in a lecture about consciousness at the Smithsonian Institution. When I suggested that we should be open to nonlocal communication between humans and animals, in view of the fact that we really have no materialistic explanations for these events, a man in the audience became quite upset. "It isn't magic, it's pheromones!" he said. Pheromones are various chemical substances secreted externally by certain animals such as ants and moths that elicit specific responses. "The dog tuned in to the pheromones of his owner," he maintained, "which were carried on the prevailing winds that blow west to east." He was right about the wind direction but wrong about pheromones: they function only between members of the same species. And even if we grant Bobby a keen sense of smell, no one knows how he could have separated his owner's scent from all the other scents that got mixed with it while wafting three-quarters of the way across the United States. Moreover, the pheromone-and-prevailing-wind hypothesis could not explain how animals return to their owners when the winds are blowing in the wrong direction.

The man's reaction was typical of people who understandably find this distant connection threatening to what they hold to be facts. In explaining animal behaviors such as these, the pure mechanists prefer to fall back on chance coincidence, predictable routine, keen senses of smell and healing, subtle cues, or wishful thinking on the part of observers who prefer magic over science. But there is a point where mechanistic explanations can become so fantastic that it seems simpler to propose an alternative connection, such as nonlocal mind.

Particularly fascinating are those cases in which the returning animal appears to be responding to the physical and emotional needs of the remote person. An example is that of an Irish soldier in World War I, whose wife and small dog, Prince, took up residence in 1914 in Hammersmith, London, while he was sent with one of the earliest contingents to the battlefields of France. After a period of service he was granted leave to visit his family, but when he returned to battle Prince was utterly disconsolate and refused all food. Then the dog disappeared. For ten days the wife tried desperately to trace him, to no avail. Finally she decided to break the news in a letter to her husband. She was astonished when she heard from him that the dog had joined him in the trenches at Armentières, under heavy bombardment. Somehow Prince had made his way through the streets of London, traversed seventy miles of English countryside, crossed the English Channel, traveled over sixty miles of French soil, and then had "smelt his master out amongst an army of half a million Englishmen and this despite the fact that the last mile or so of intervening ground was reeking with bursting shells, many of them charged with tear-gas!"[28]

Napoleon, in his Italian campaign of 1796, strolled one night through a bloody battlefield following victory. Suddenly a dog leaped from the body of its dead master toward Napoleon, then retreated to lick the hand of the dead man, howling pitifully. The dog repeated the action over and over—rushing toward Napoleon and retreating to his slain master. This unbreakable link between the dog and the dead soldier moved Napoleon deeply. "No incident on any field of battle," he wrote, "ever produced so deep an impression on me. I involuntarily contemplated the scene. This man, thought I, had friends in his camp, or in his company; and now he lies forsaken by all except his dog! What a lesson nature here presents through the medium of an animal!"[29]

Wild animals also respond to human need. Eighty-two-year-old Rachel Flynn was taking her customary walk on Cape Cod one day in 1980, when she fell off a thirty-foot cliff onto a lonely beach. Too badly hurt to move, she thought she would die. Lying trapped between boulders, she saw a seagull hovering over her. She remembered that she and her sister had regularly fed a gull, which they named Nancy, at their home. Could this be the same bird? Acting on a long shot, Miss Flynn cried, "For God's sake, Nancy, get help." The gull flew off toward her home, a mile away, where June, her sister, was working in the kitchen. June described later that she was irritated by a seagull that began tapping on the windowpane with its beak and flapping its wings, "making more noise than a wild turkey." She could not shoo it away. After fifteen minutes of the gull's frantic behavior, it occurred to June that the wild bird—it was not a pet—might be trying to tell her something. Going outside, she followed the bird as it flew ahead. It stopped occasionally, as if to make sure she was following. The seagull alighted on the cliff over which Miss Flynn had fallen. June summoned an ambulance, which rescued her bruised and helpless sister.[30]

British biologist Rupert Sheldrake's 1997 survey of pet owners in northwest England found that 53 percent of dog owners and 33 percent of cat owners thought their pet responded to their thoughts or silent commands, and similar percentages believed their pet was sometimes telepathic with them. Sixty-nine percent of dog owners and 48 percent of cat owners thought their pets knew when they were going out before they showed any physical signs of doing so, and similar percentages believed their pet knew when some member of the household was on their way home.[31]

After publicly announcing his interest in this area of research in 1992, Sheldrake was flooded with letters from pet owners such as Louise Gavit of Morrow, Georgia, who wrote:[32]

> In our case, there is no habit or schedule to my comings and goings, yet my husband tells me (and from past experience with two cats and one dog who did the same) our dog always responds to my coming home. In fact he seems to respond to my *intention and action* to come home. As near as I can measure my movements in comparison to the actions taken by the dog, his responses to my mental and physical actions are as follows: as I leave the place I have been and walk to my car with the intent to come home, our dog BJ will awaken from sleep, move to the door, lie down on the floor near the door, and point his nose toward the door. There he waits. As I near the drive, he becomes more alert and begins to pace and show excitement the nearer I move to home. He is always there to poke his nose through the crack, in greeting, as I open the door. This sensing seems to be unlimited by distance. He does not seem to respond at all to my leaving one place and moving to another, his response seems to be apparent at the time when I form the thought to return home, and take the action to walk toward my car to come home.

Sheldrake suggested to Gavit that she try coming home by unusual means, such as by being driven by someone else in an unfamiliar car. She responded that this seemed to make no difference:

> My method of travel is irregular: using my own car, my husband's car, a truck, or any number of cars driven by strangers to BJ, or walking. Somehow BJ responds to my thought/action just the same. Even when BJ has seen my car still inside the garage, which is located in the basement of my home, he responds.

Sheldrake has documented the responses of pets through video recordings, which allows an objective assessment of the animal's behavior by individuals other than a household member. He can tell whether the pet is acting routinely by varying the time that the owner departs from work and arrives home. And he can tell if the pet is picking up the expectations of the person waiting at home by keeping that person in the dark about the departure and arrival times as well. Sheldrake's data shows that many pets cannot be thrown off by manipulating these factors. They seem nonlocally linked to their owners, as if they are reading their thoughts.

THE MYSTERY OF HOMING PIGEONS

A bit more about Rupert Sheldrake, who is one of the most creative scientists I have ever met. He was a scholar of Clare College, Cambridge, where he read natural sciences, took a Ph.D. in biochemistry, and was Director of Studies in biochemistry and cell biology from 1967 through 1973. He also studied philosophy at Harvard University, and from 1974 through 1978 he was a member of the staff of the International Crops Research Institute for the Semi-Arid Tropics at Hyderabad, India.

I first met Sheldrake in the 1980s, when our mutual interests in the nature of consciousness brought us together at conferences. On one occasion he visited my wife, Barbara, and me in Dallas, Texas, where he had been invited to lecture. One memorable evening we dined together at an Indian restaurant, in remembrance of his work on food crops for the people of India. Sheldrake had just published his landmark book, *A New Science of Life*,[33] and I was in the process of writing *Recovering the Soul*. I was struggling at the time with what to call this factor of consciousness, which escapes its confinement to the brain and body and behaves on occasion as if there is no space and no time. "What do you think about the term *nonlocal mind?*" I asked hesitantly over samozas and curried rice. Sheldrake was the perfect scientist to ask; he was familiar with the concept of nonlocality in physics and had considered how it might apply to living things. After pondering the question he said, "I think it is perhaps the best term I have heard." His opinion encouraged me to introduce the term *nonlocal mind* in my 1989 book.[34] If a scientist of his caliber found it useful, then, I suspected, others would, too.

Since then I have discussed nonlocal mind on many occasions with Sheldrake. As a biologist, he feels deep personal connections with plants, animals, and birds, and he believes that experiments with them can shed light on the nonlocal nature of the psyche. It did not surprise me to discover that Sheldrake has been a pigeon fancier since childhood. Pigeons were also a passion of Charles Darwin, who kept a wide range of breeds. Darwin puzzled over how pigeons return home across great distances, often against great odds, and he even published a paper on the subject in 1873 in *Nature.* He suggested that they do it by a kind of "dead reckoning," somehow registering and computing all the twists and turns of the outward journey, which they remember on their flight back home. Sheldrake has analyzed this and all the other attempts of scientists to explain how pigeons home in his challenging book *Seven Experiments That Could Change the World.*[35] Even though this question has been investigated extensively for more than a century, no one knows how the birds do it.

In the case of migrant birds, the most popular hypothesis is that they orient by the stars through what is called an "inherited spatio-temporal vector-navigation program"—an impressive-sounding term, states Sheldrake, that merely restates the problem rather than solving it. It seems unlikely that racing pigeons, who have been shut up in baskets and transported "hundreds of miles in trains, trucks, ships, or aeroplanes, subject to many twists and turns, can continuously compute their homeward direction with the highest precision."[36] The idea seems also to have been refuted by one researcher, who heavily anesthetized pigeons before they were transported to the release site. On recovering, they were still able to find their way home.

Attempts to deliberately disorient pigeons by transporting them in crazily rotating drums also have failed. In addition, the semicircular canals in the ear, which are involved in balance, acceleration, and rotation, have been surgically cut, after which the pigeons perform as well as controls.

Do pigeons rely on familiar landmarks to find their way back? Perhaps to some extent, but this doesn't explain how pigeons can home across hundreds of miles of territory they have never seen or how they can home over vast stretches of ocean with no landmarks or during dense fog or at night when landmarks are obscured.

Is vision required? Pigeons have even been fitted with frosted-glass contact lenses so they can't see, and they return home as successfully as pigeons fitted with clear contact lenses. A favorite hypothesis is that pigeons rely on the sun's elevation and arc, aided by some sort of internal chronometer. But "following the sun" fails to explain how they home with frosted contact lenses on, and at night, and in heavily clouded conditions.

Do pigeons "smell" their way home? If smell is required, they should not be able to home when the wind is blowing toward rather than away from their destination. To further test the role of smell, the air sacs in their olfactory organ have been punctured by needles, and they still home normally. The nasal cavities have also been stuffed with wax, with no detriment to their homing skills.

Do pigeons have a highly developed magnetic sense? At best, this could afford only a partial explanation for homing, Sheldrake asserts. The earth's magnetic field is far from uniform but varies from place to place depending on the underlying rocks. These magnetic anomalies may be a few yards or hundreds of miles across. In addition, the earth's magnetic field is not constant but fluctuates daily. Even if pigeons could somehow correct for these anomalies, the earth's magnetic field would give no information on movements in an east-west direction.

It is widely assumed, however, that scientists have proved that birds migrate and pigeons home according to an internal magnetic sense. This is not true. Sheldrake has examined all the experiments done to evaluate this possibility. They give conflicting data and cannot be replicated by the scientists who initially performed them. "The magnetic hypothesis was the last seemingly viable attempt to find a [physical] mechanism for homing," Sheldrake concludes. "Many have clung to it with the tenacity of drowning men clutching straws. Now this hypothesis too has sunk," he asserts. "Among professional researchers, the currently conventional view is that pigeon homing depends on a complex series of 'backup systems'; or that it is 'multifactorial,' involving subtle combinations of mechanisms. . . . But these scientific-sounding phrases merely disguise a profound ignorance."[37]

The answer to the puzzle may prove to be a nonlocal one. "I propose that the sense of direction of homing pigeons depends on something rather like an invisible elastic band connecting them to their home, and drawing them back toward it," says Sheldrake. "When they are taken away, this band is stretched. If on their return flight they overshoot their home, as some of the pigeons flying with frosted-glass contact lenses did, this connection serves to pull them back again."

How does this interconnection work? Sheldrake says he does not know. "It might be related to the nonlocal connections implied by modern quantum physics. . . . Perhaps the link between the pigeon and its home rests upon such nonlocal quantum phenomena," he proposes. "Perhaps it does not, but rather depends on some other kind of field or interconnection not yet recognized by physics. I simply leave this question open." Sheldrake notes that in the field of modern dynamics, mathematicians describe how systems move within a field-space toward "attractors." "Perhaps a pigeon could be modeled as a physical body moving within a vector-field toward such an attractor, representing its home or goal."[38]

If consciousness is nonlocal, another possibility exists. If one pigeon knows the location of the pigeon loft at home, this awareness might be available nonlocally to other pigeons. This might explain the puzzling event that happened in June 1995 during a pigeon race in Europe.

A female racing pigeon belonging to David Dougal, of Northumberland, England, was supposed to have flown to his home from Veau Vois, France. Instead she headed southwest to the North African coast, settling into a pigeon loft in Morocco owned by Essoli Mohamed. In October, a second bird, a nephew of the first, also set off from Dougal's home and flew the sixteen hundred miles to join his aunt. "I couldn't believe it when I got another letter from Essoli," said Dougal. "When the first bird went missing we weren't really too surprised because it was awful weather for the race and that affects the bird's homing instincts. But there is no explanation we can come up with for the nephew. He was only a few weeks old and had just learnt to fly." Dougal, astonished that the second bird was able to single out his aunt's adopted home from all the other pigeon lofts in the world, let the birds remain in Morocco.[39]

"Man has the capacity to love, not just his own species, but life in all its shapes and forms," wrote anthropologist and poet Loren Eiseley. "This empathy with the interknit web of life is the highest spiritual expression I know of."[40] Perhaps empathy and love exist not only between humans but between other creatures as well. If so, the interknit web of life may be wider than we commonly imagine. It may be boundless, limitless, nonlocal. It may even involve chickens.

CAN THE INTENTIONS OF ANIMALS CHANGE THE WORLD NONLOCALLY?

Scientists interested in the nature of consciousness sometimes wonder whether animals, like humans, can influence the world nonlocally. Among them is French researcher René Peoc'h. At the University of Nantes he examined the ability of baby chicks to influence random, physical events. It is well known that birds, after hatching, readily imprint or adopt as their mother the first close moving object they see. Peoc'h and his colleagues conditioned chicks to adopt a small robot called Tychoscope I as their mother by placing them for one hour alone in the presence of the moving robot every day for six days following their birth. The robot was a small, self-propelled device designed to move about on a level surface in successive segments of random length and orientation. The path inscribed by Tychoscope I could be graphically recorded and analyzed by a computer. After the conditioning, the chicks were placed singly in a transparent cage, from which they could see the robot moving around on the floor. The objective of the experiment was to see whether Tychoscope I would continue to move at random or if it would migrate toward the chicks who had become bonded with it. It was discovered that the robot spent two and a half times longer on the half of the surface closer to the chicks, compared to its motion when the cage was empty. Using chicks that had *not* been conditioned to adopt the robot as their mother, Tychoscope I moved about randomly when in their presence. Therefore it seemed that the physical motion of the robot depended more on the conditioning—the feelings of bondedness and closeness? of love?—of the chicks than on the physical processes driving the robot. There was less than one chance in a thousand that these results could have been due to chance.[41]

Based on these findings, Peoc'h and his colleagues undertook a new series of experiments, published in 1995. They developed a second-generation robot, Tychoscope II. This robot was remotely controlled to move randomly by a computer, which recorded and analyzed all its movements. In contrast to the first series of experiments, the chicks in these studies had not previously been exposed to the robot. They had, however, been raised in the dark. The goal was to see whether groups of these young chicks could affect the path of Tychoscope II when it was carrying a lighted candle. Chicks do not like to be in the dark during the day. Could they "pull" the light-bearing robot toward them in order to have more light?

Eighty groups of fifteen chicks each were tested. In 71 percent of the cases, the robot spent excessive time in the vicinity of the chicks. In the absence of the chicks, it followed random trajectories. There was less than one chance in a hundred that these effects were explainable by chance.[42]

IS THE WORLD "THOUGHT HUNGRY"?

The idea of nonlocal mind, as we've seen, usually triggers the image of an individual sending thoughts or intentions outward into the world, where they make something happen. For example, we speak of *sending* prayer outward (and usually upward) to the person in need. These images suggest that we are the active agent and that the recipient is passive. But evidence suggests that the world, not us, may be doing the work. It may be actively *absorbing* our thoughts while we are doing nothing at all, including not thinking or intending that things change. It's as if the world is "thought hungry," that it *wants* what we think and feel.

While I was exchanging letters with San Diego physician-author Dennis Gersten about this concept, an event happened that seemed to affirm this possibility. Gersten related an incident in the life of the Indian spiritual master Sathya Sai Baba. On one occasion he was presenting saris as gifts to women who were his followers. One of his assistants noticed that one unopened box of saris was dripping liquid from a corner, and he called Sai Baba's attention to it. Sai Baba explained that that particular box of saris had not been of high enough quality to be given away. The saris were grieving that they were not worthy of being considered as gifts, and the tears they were shedding were seeping from the bottom of the box.[43]

One morning I was excerpting this charming story from Gersten's letter for an essay I was writing. I turned off my computer to attend to another task. When I turned it back on, the screen brightened, then instantly became black, and I experienced that feeling of dread known to all writers who fear their material has evaporated into virtual reality. Then, in the center of the screen, an image appeared that I had never seen before—a computer icon with an unhappy face and a downturned smile. My wife, Barbara, who was in my office at the time, glanced at the icon and remarked, "Your computer isn't very happy." It was as if the computer had caught the mood of the sorrowful saris, whose description I had just finished recording in its inner workings, and was crying along with them. The computer had crashed. I could not erase the unhappy face from the screen and had to call the computer help line for advice about how to proceed.

Skeptics will say that all computers crash sooner or later and that this chain of events was purely coincidental. Philosophers have a special term, the *pathetic fallacy,* to express their disdain about attributing human feelings to inanimate things. When poets speak of an "angry" sea or "playful" winds or when physicians refer to a "stubborn" case of high blood pressure, they are unjustified and are distorting what nature is really like. But occasional scientists have taken quite seriously the possibility that things can feel. Niels Bohr, one of the patriarchs of modern physics, observed, "I am absolutely prepared to talk of the spiritual life of an electronic computer; to state that it is reflecting or that it is in a bad mood. . . ."[44]

Can computers and electronic gadgets absorb our moods? The answers researchers are giving to this question is another line of evidence favoring nonlocal mind.

PRINCETON AND THE RANDOM EVENT GENERATOR

While surveying the research underlying nonlocal mind, I visited the most famous laboratory in the world currently investigating this area—PEAR, the Princeton Engineering Anomalies Research facility. The PEAR lab was founded by Robert G. Jahn, Princeton's former dean of engineering. In the 1970s a graduate student of Jahn's proposed a research study to examine whether or not an individual might be able to affect the function of sensitive electronic devices. Jahn, usually a skeptic about such matters, agreed. To his amazement, the experiment worked. This set in motion further research out of which PEAR grew. Over the past two decades the PEAR researchers have observed *millions* of trials in which individuals try mentally to influence some physical phenomenon at a distance—the largest database on the subject in the history of science.

I was surprised during my visit to the PEAR facility. I'd expected a swarm of white-coated scientists and banks of computers in a sterile, laboratory setting. Instead, I felt as if I were entering a homey apartment with soft couches, stuffed animals, attractive art, and some of the friendliest people I've ever met. Although there were experimental gadgets in every nook, they were there to serve, not dominate. I participated in an experiment and watched several others. I had fun.

One of the PEAR team's favorite devices is a computerlike gadget called a random event generator, or REG. These instruments are often called electronic coin flippers because their output is truly random. If the REG is allowed to run over an extended period of time—hours or days—it will churn out an equal number of heads or tails, ones or zeros, pluses or minuses. For more than twenty years now, the PEAR researchers have explored whether or not individuals can mentally influence the randomness of an REG's performance.[45] The "operator" sits in front of a computer screen and watches a line wander aimlessly above and below a certain point, which reflects the REG's output of pluses and minuses. Because it randomly produces equal numbers of each, the line spends equal time above as below the midpoint. After stating the direction in which they intend to "push" the computer's output, the would-be influencer then puts his or her mind to the task of making the computer "go high" and produce more pluses, or "go low" and churn out more minuses. The results, from millions of trials by scores of operators, indicate that the REG regularly responds to the desires of the operators, suggesting that the mind can literally change the external world.

Next, the PEAR team turned their attention to distance. If the operator is far away, they wondered, does the effect persist? The PEAR team has separated the operators and the computers as far as possible, positioning them on opposite sides of the earth. The effects of mental intentions came through as robustly and as reliably as ever. Distance, they proved, is irrelevant for the power of the mind to influence its surroundings.[46]

Next, what about time? The PEAR team has shown that operators do not have to focus on the REG at the moment it is running in order to influence it. Even if they concentrate on its performance *before* it has actually run, they are able to manipulate its output in the desired direction, suggesting that the mind can act into the future. Can the mind affect past events? To test this possibility, the machine is allowed to run and the results are electronically recorded but not actually observed. Because no one has influenced it, the recorded output should be random. But if the operator attempts to influence the REG's output hours or days *after* it has run, and the prerecorded output is *then* examined, it is found to be skewed in the direction of the operator's intent. This suggests that the mind can reach back into the past and influence events that presumably have already happened.[47]

Are individuals or groups better able to influence the output of REGs? The highest scores observed at the PEAR lab occur when two people of opposite sex, who are emotionally bonded, try to influence the REG together.[48]

What if lots of people try to influence the physical device? Is more better? This question is commonly asked by people who pray. Is group prayer more effective than when people pray singly? Seeking an answer, the PEAR team developed a portable REG that could be taken to group settings. They took the REG to ten separate conferences and stationed the REG in a nearby room and collected data at various moments during various sessions and presentations, usually without the participants' awareness that the REG was running. They found that when the participants shared a high degree of attention or experienced a common emotion, the REG tended to depart from its expected random output. After performing sophisticated statistical analyses, they found that there were less than two chances in ten thousand that the results could be explained by chance. The PEAR experimenters hypothesize that there is a collective consciousness among people that forms a kind of field—thus the term *field* REG, which has been given to these studies.[49]

In the experiments involving single individuals sitting in front of the REG, successful operators commonly say they somehow "become one" with the instrument. In fact, the most successful individual actually described falling in love with the machine. Even when large groups influence the REG, as in the meetings and conferences, an emotional process also appears to be taking place. The PEAR team analyzed logbook notes and anecdotal reports from members of the ten groups they tested. The most striking REG effects took place when an emotional coherence or togetherness penetrated the groups—what the PEAR team has begun to call "resonance."

THE EXPERIENCE OF BEING STARED AT

One of the most challenging questions about nonlocal effects of the mind involves whether or not an experimenter's desires can affect the experiment itself. Most scientists cringe at this possibility. What one *thinks* about the outcome of an experiment is supposed to be irrelevant. If different experimenters follow the same methods, the outcome of an experiment should be the same, no matter what they believe. Insights into these assumptions have come from an unexpected quarter —the common experience of being stared at.

Almost everyone has had the experience of feeling like they are being stared at in a restaurant or at a traffic light and has looked around to see someone gazing at them. But can we really tell when someone is staring at us? This question has been asked by researchers for over a hundred years, with variable findings.[50] Research took an interesting turn in 1983 with the work of Linda Williams, an Australian undergraduate student at the University of Adelaide. She stationed the starers and recipients in separate rooms sixty feet apart. The recipient was viewed by the starer on a closed-circuit television, which prevented the use of subtle cues for communication. The starer viewed the subject on the screen for twelve seconds, or the screen was blank, according to a random sequence. The recipient was told by an electronic bleep when each new twelve-second trial began, and the recipient was told to indicate whether or not he or she felt stared at. In twenty-eight subjects, a small but statistically significant positive effect was found.[51]

In 1990, researchers William Braud, Donna Shafer, and Sperry Andrews at the Mind Science Foundation in San Antonio, Texas, followed Williams's direction with three experiments in which individuals stared intently at a closed-circuit television image of another person in a distant room. Unlike all previous studies, the recipients of the stares were not asked to indicate when they felt they were being stared at. Instead, a computer measured their electrodermal response, which is the capacity of the skin to conduct an electrical current, and which is an indicator of physiological arousal. Numerous staring periods were randomly interspersed with nonstaring, control periods, and the recipient was not informed about the timing and scheduling of the various periods. The electrodermal activity levels were significantly different for staring versus nonstaring periods. These experiments suggest that it is possible to perturb the physiology of a distant person nonlocally, merely by looking at *an image* of the person, outside their awareness. These experiments were replicated by researchers Marilyn Schlitz and Stephen LaBerge in 1997 in a rigorous study at the Cognitive Sciences Laboratory at Science Applications International Corporation in Palo Alto, California.[52]

Then the staring studies took a fascinating turn.

One of England's best-known skeptics of nonlocal mental phenomena is Richard Wiseman of the University of Hertfordshire. Wiseman carried out staring studies in 1994 and found no evidence of a nonlocal effect, in contrast to Dr. Marilyn Schlitz, whose above staring experiments were successful. Why the difference? There were several possible explanations. Schlitz's studies may have contained an undetected glitch that was absent from Wiseman's experiments, or vice versa. Or perhaps Schlitz had worked with participants who were more psychically gifted than Wiseman's. Or Schlitz may have been more skilled at eliciting nonlocal mental abilities from her participants than was Wiseman. Perhaps one experimenter cheated or faked the data.

But another possibility struck both Schlitz and Wiseman: Wiseman was convinced that nonlocal mental effects do not exist, while Schlitz was convinced that they are real. Could their preexisting beliefs have operated nonlocally to influence the outcome of their experiments?

In search of an answer, Schlitz and Wiseman decided to collaborate in a staring study.[53] Schlitz traveled to England, where she and Wiseman acted as separate experimenters for two different sets of trials. The experiments were carried out at the same time and in the same location—Wiseman's laboratory at the University of Hertfordshire. They used the same equipment, drew subjects from the same subject pool, and employed exactly the same rigorous methods and procedures they had used previously. The subjects were in a distant part of the building, with their skin resistance being measured and recorded. A video image of them was displayed to the starers, who were asked to stare at it according to a random sequence. The only real difference between the two trials was that one was carried out by Schlitz and the other by Wiseman.

Wiseman's subjects showed no change in their physiology when they were and were not being stared at. Schlitz's subjects, in contrast, were significantly more activated when stared at, even though they were unaware when they were being observed.

It is unlikely that the different results were due to an experimental artifact. The same methods were followed, so an experimental mistake should have affected Wiseman's and Schlitz's trials equally. Could the subjects have cheated? The conditions were so stringently controlled that this is virtually impossible. Could Wiseman or Schlitz have engaged in fraud? The experiments were designed to make this highly unlikely, and no cheating was discovered during the running of the studies or the analysis of the data. Could one set of subjects been more "psychically gifted" than the other? They came from the same pool, so this is not likely. Could Schlitz have been better at eliciting greater nonlocal performance from her subjects than Wiseman? Could the subjects have picked up on the differing beliefs of the experimenters through their interactions with them and allowed the experimenters' beliefs to affect them? This possibility is being evaluated by analyzing videotapes of how Schlitz and Wiseman interacted with the subjects. Or did Wiseman and Schlitz, through their differing beliefs and intentions, nonlocally steer the outcome of the experiment in opposite directions?

If our intentions, beliefs, and wishes can affect the outcome of scientific experiments, many scientists believe this would mean the end of science. This fear is probably overwrought. Some physical phenomena are so robust they would likely be resistant to being pushed around by our intentions—the paths of the planets, the behavior of volcanoes, or the drift of continents. Other natural phenomena, such as subatomic or quantum processes that control computers, might be more sensitive to the nonlocal caresses of the psyche.

ARE WE ZOMBIES?

Most laypeople, I've discovered, are often puzzled about why research studies such as these continue. They assume that scientists by now must surely have figured out the basic relationship between the mind and the brain. They don't realize how appalling our ignorance really is.

In the various "consciousness conferences" that are springing up everywhere these days, scientists often engage in white-knuckled arguments about whether or not consciousness *exists,* let alone how it works. This state of affairs highlights our primitive understanding of the mind. Daniel Dennett, a distinguished commentator on the relationship between the mind and brain and author of *Consciousness Explained,* has concluded that there is really no such thing as consciousness—we are all zombies and automatons who, through some freak of nature, *think* we're conscious. A similar position is taken by Francis Crick, who shared a Nobel Prize in 1953 for the discovery of the structure of DNA. In his 1994 book, *The Astonishing Hypothesis,* he wrote that "your joys and your sorrows, your memories and your ambitions, your sense of personal identity and free will, are in fact no more than the behavior of a vast assembly of nerve cells and their associated molecules."[54]

Laypeople often scratch their heads in amazement at how a scientist could adopt such a position. If Crick is correct, then he, like everybody else, has no free will because he is compelled by his own "vast assembly of nerve cells and their associated molecules" to say what he says. He did not freely arrive at his conclusions by a carefully considered weighing of evidence; rather, his brain *made* him think the way he thinks; he had no choice in the matter. Why, then, should we believe him? If he is "on automatic," as he contends, why doesn't he close his lab door, turn off his instruments, shut his books, and go home? Why lobby congressional representatives to fund his research and approve his grant proposals, since they, like everyone else, have no choice? Can people who hold this point of view actually believe it? How can they participate in science, which requires freedom of inquiry, if they think they are not free to inquire?

If these theorists are correct, then mind-body interactions are really only brain-body interactions, and we are deluding ourselves if we think that we can intentionally affect ourselves or others through our thoughts, wishes, or prayers. But there is no reason to close the books on the role of consciousness in health. Even if the brain were capable of actually producing consciousness, for which there is no evidence, no one has a clue about how the *contents* of the mind—specific thoughts—would come about as a result. "No," Einstein once said in response to this idea, "this trick won't work. . . . How on earth are you ever going to explain in terms of chemistry and physics so important a biological phenomenon as first love?" His contemporary, Niels Bohr, whose name is synonymous with modern physics, agreed that physics cannot shed much light on the mind. "We can admittedly find nothing in physics or chemistry," he said, "that has even a remote bearing on consciousness."[55]

In response, proponents of the mind-equals-brain perspective assure us that when we know enough—when we are able to explain the brain's deepest workings in terms of the behavior of subatomic particles and quantum physics, for example—*then* we'll see that the operations of consciousness are totally physical in nature. Experts disagree. As Nobel physicist Eugene Wigner stated, "Quantum mechanics is passing the buck. . . . We have at present not even the vaguest idea how to connect the physico-chemical processes with the state of mind. . . ."[56] Further, he says,

> The most important phenomena not dealt with by our physical theories are those of life and consciousness. . . . Even if the physical theories could completely describe the motions of the atoms in our bodies, they would not give a picture of the content of our consciousness, they would not tell us whether we experience pain or pleasure, whether we are thinking of prime numbers or of our granddaughters. This fact is, in my opinion, the most obvious but also the most convincing evidence that life and consciousness are outside the area of present day physics.[57]

Physicist and astronomer David Darling also considers the brain-equals-mind point of view a dead end. He writes,

> Working under this assumption [that consciousness is produced by the brain], a growing number of scientists are now busily rummaging around in the brain trying to explain how the trick of consciousness is done. Researchers of the stature of Francis Crick, Daniel Dennett, Gerald Edelman, and Roger Penrose have recently come forward with a range of ingenious theories. All purport to explain, in one way or another, consciousness as an epiphenomenon of physical and chemical processes taking place in the brain—and all fail utterly. They fail not because their models are insufficiently accurate or detailed, but because they are trying to do what is, from the outset, impossible.
>
> The truth is that *no* account of what goes on at the mechanistic level of the brain can shed any light whatsoever on why consciousness exists. No theory can explain why the brain shouldn't work exactly as it does, yet without giving rise to the feeling we all have of "what it is like to be." And there is, I believe, a very simple reason for this. The brain does not *produce* consciousness at all, any more than a television set creates the programs that appear on its screen. On the contrary, the brain *filters* and *restricts* consciousness, just as our senses limit the totality of experience to which we might otherwise have access.[58]

Darling is expressing a point of view that is central to Era III—that mind is *fundamental* in the world but is not the exclusive property of the physical body. Far from being a fringe idea, this perspective has been embraced by scientists of the highest caliber, such as Nobel laureate in biology George Wald: "Mind," he says, "rather than emerging as a late outgrowth in the evolution of life, has existed always . . . , the source and condition of physical reality."[59]

One of the most common objections against the nonlocal, distant operations of the mind is that these events violate the laws of nature. Eminent scientists say otherwise. Physicist Gerald Feinberg has stated, "If such phenomena indeed occur, no change in the fundamental equations of physics would be needed to describe them." And psychologist Paul Meel and philosopher of science Michael Scriven have pointed out that this objection rests on two highly questionable assumptions: that current scientific knowledge is complete and that distant mental effects conflict with it. Physicist O. Costa de Beauregard states, "Today's physics allows for the existence of the so-called 'paranormal' phenomena of telepathy, precognition, and psychokinesis. . . . The whole concept of 'non-locality' in contemporary physics requires this possibility."[60]

Among the prestigious scientists who have supported an open hearing for nonlocal mind is physicist Henry Margenau, Eugene Higgins Professor Emeritus of Physics and Natural Philosophy at Yale University. Professor Margenau taught physics for forty-one years at Yale and was a staff member of the prestigious Institute for Advanced Study at Princeton University. For a half century before his recent death at age ninety-six, Margenau made central contributions to modern physics. He was also interested in nonlocal phenomena such as we've examined, including events that are studied by parapsychologists. I quote the following comments of Professor Margenau at length because they neatly sum up our current situation, suggesting why we should be open to a nonlocal view of reality.

> It appears to be a matter of common sense to any scientifically trained person today that ESP (telepathy, clairvoyance, precognition) is impossible, since such phenomena—if they existed—would violate known and proven scientific laws. On this basis we can confidently predict that reports of occurrences of this kind are due to poor observation, bad experimental design, and outright chicanery. . . .
>
> This is the attitude of many modern scientists. . . .
>
> However, a question can be raised as to exactly what scientific laws would be violated by the occurrence of ESP. . . .
>
> Does ESP violate the canon against "action at a distance"? Perhaps it would if there were such a universal principle. There are current, at present, respectable conjectures among physicists who introduce massless fields in which phenomena can be transmitted instantly. In quantum mechanics, a debate is raging about non-locality of interactions; the term is a high-brow version of action-at-a-distance. . . . ESP is no stranger than some of the discussions in this field.
>
> Strangely, it does not seem possible to find the scientific laws or principles violated by the existence of ESP. We *can* find contradictions between ESP and our culturally accepted view of reality, but not—as many of us have believed—between ESP and the scientific laws that have been so laboriously developed. Unless we find such contradictions, it may be advisable to look more carefully at reports of these strange and uncomfortable phenomena which come to us from trained scientists and fulfill the basic rules of scientific research. We believe that the number of these high quality reports is already considerable and increasing.[61]

In view of our ignorance about the nature and origin of consciousness, the wisest stance, it would seem, is one of openness. Where the workings of consciousness are concerned, we need to keep our options open. Astronomer Carl Sagan, although no fan of nonlocal mind, nonetheless advocated a position I heartily endorse, which I believe will serve us well as we enter discussions of Era III. In his UCLA commencement speech on June 14, 1991, he said, "It is the responsibility of scientists never to suppress knowledge, no matter how awkward that knowledge is, no matter how it may bother those in power. We are not smart enough to decide which pieces of knowledge are permissible and which are not. . . ."

THE EMERGING PICTURE OF CONSCIOUSNESS

No one can say what the final picture of consciousness will be. But as researchers Dean I. Radin, Janine M. Rebman, and Maikwe P. Cross suggest, the new model of the mind will almost certainly show us that:

- Consciousness is *nonlocal.* It extends beyond the individual. It cannot be confined to specific points in space, such as brains or bodies, or specific points in time, such as the present moment.
- Consciousness is an ordering principle. It can insert information into disorganized or random systems and create higher degrees of order.
- Consciousness is not the same as awareness. The ordering power of consciousness can occur completely outside awareness, such as in dreams.
- Both individual and group consciousness can insert order or information into the world, and can extract information from the world as well.
- Coherence among individuals is important in the ordering power of consciousness. Coherence may be expressed as love, empathy, caring, unity, oneness, and connectedness.
- Consciousness can affect humans and nonhumans alike. Even inanimate objects can "resonate" with and respond to human consciousness.[62]

When Sir Isaac Newton began trying to make sense of his world in the 1600s, he invented new terms to help him do the job. One of these, *universal gravity,* was so controversial that his colleagues accused him of selling out to mysticism. Newton, they charged, had no explanation for how this invisible force worked, and he could not clarify how solid objects could obey the dictates of invisible gravity even if they wanted. Undeterred, Newton stuck to his observations and to his mathematical calculations. The data, he claimed, made it *mandatory* to propose the existence of gravity; how his opponents thought the world *ought* to behave was irrelevant. Newton's colleagues eventually got over their intellectual indigestion about gravity, and today gravity passes for common sense. But this is not because we understand it. We *still* don't know how gravity works; we have merely gotten used to it.

So it is with nonlocal mind. We are required to propose its existence because of scientific data and our own experiences. Skeptics may moan, bellow, howl, and whine, as they did against gravity, but nonlocal mind is an idea whose time has come. And years hence, when the smoke of controversy clears, we shall regard nonlocal mind as obvious, natural, and *right*—as gravity.

CHAPTER 3

DREAMS, PRAYER, AND THE ZONE: HOW WE EXPERIENCE NONLOCALITY

The brain breathes mind like the lungs breathe air.

HUSTON SMITH
Forgotten Truth

Nonlocal mind often bursts into our lives spontaneously, unannounced, "ready or not." Its sudden appearance can be a shock.

Marilyn Ferguson, author of *The Aquarian Conspiracy* and editor-publisher of *Brain/Mind Bulletin,* suddenly heard her name spoken aloud, out of the blue.[1] She recounts,

> Naturally, I was jarred. During the week just passed I had been reading a book on extrasensory perception, a topic I had always avoided. Briefly I wondered if the voice meant that someone close to me was endangered or dying. And then I chastised myself for an overactive imagination.

> When I returned two hours later, my husband was waiting to tell me that my apparently healthy father, 1,000 miles away, had just suffered a fatal heart attack. The next day, when I found that two other family members had had premonitions of his death, it struck me that these phenomena must be almost commonplace.
>
> If so, I reasoned, our understanding of time and space must be incomplete.[2]

Ferguson's initial reaction—to chastise herself for an overactive imagination—is a response many of us have to a nonlocal experience. As we grow up, we learn to think locally. By the time we're adults, we "know" that our mind is confined to our brain and that all forms of communication must be mediated by our physical senses. Experts on every hand remind us that nonlocal events are "funny coincidences," at best, and hallucinations experienced by weak-minded individuals, at worst.

These criticisms are off the mark. Surveys by the University of Chicago's National Opinions Research Council (NORC) find that "people who have tasted the paranormal, whether they accept it intellectually or not, are anything but religious nuts or psychiatric cases. They are, for the most part, ordinary Americans, somewhat *above* the norm in education and intelligence and somewhat *less* than average in religious involvement." Sociologist Andrew Greeley of NORC tested people who had profoundly mystical experiences, such as being bathed in white light. When these persons were subjected to standard tests measuring psychological well-being, the mystics scored *at the top.* University of Chicago psychologist Norman Bradburn, who developed the test, said that no other factor had ever been found to correlate so highly with psychological balance as did mystical experience.[3]

How common are nonlocal experiences? Researchers at NORC, who have polled Americans on their inner life since 1973, find that around two-thirds say they have experienced some sort of extrasensory perception, and nearly half report "contact with the dead." Around 30 percent have had "visions," two-thirds have experienced *déjà vu,* and one-third have seen things at a distance.[4]

As Ferguson's experience shows, it can be difficult to make sense of our lives unless we acknowledge that our mind operates nonlocally. If we accept nonlocal experiences as a natural part of the world, more things fall into place, make sense, and have meaning.

Consider what happened one evening to college student Emily Wright Jaeger. She was sitting in her living room studying, at the time of day when many people read the daily newspaper. All of a sudden the suggestion came into her mind, "You are dying." This made no sense to her, because she was healthy and well. Then she began to feel incapacitated—no energy, difficulty breathing, and an inability to use her legs. A deeply spiritual woman, Jaeger began to focus on the power of God, as taught in her religion. She began to pray, and she continued praying until she felt "free." Her feeling of vitality returned, and she returned to her studies.

When she returned to her office the next day, her fellow employees were shocked to see her. They behaved as if they were seeing a ghost. They revealed that they had read her name in the obituary column in the newspaper the day before. Not realizing the name belonged to a different person, her friends throughout the area were collectively believing she was dead. "I realized," she said, "that I had become mesmerized unknowingly by the thoughts of those who were ignorantly believing I was the woman in the obituary column."[5]

Jaeger had a place in her worldview for nonlocal mind. It allowed her to make sense of an otherwise upsetting, chaotic experience. If she had, say, gone to a physician who was not grounded in Era III medicine and had complained about "no energy, difficulty breathing, and inability to use my legs," and her evaluation had shown no physical problem, she almost certainly would have been diagnosed as neurotic, hypochondriacal, or worse. She might have been referred to a psychiatrist or placed on a medication to help her "gain control." It is not difficult to see why the failure to make a place for nonlocal experiences has resulted in much medical mischief.

THE ZONE

Nonlocal experiences are particularly common in sports. Making a place for nonlocal mind helps athletes understand what they call "the zone"—those enchanted moments when performance soars and the sense of the individual self is transcended.

Bill Russell is one of the greatest basketball players of all time. In 1956 he won an Olympic gold medal as a member of the U.S. basketball team, and a career with the Boston Celtics followed that resulted in eleven championships in thirteen years, during which he won the Most Valuable Player award five times.[6]

Russell dates the beginning of his greatness to a "mystical revelation" he experienced at age sixteen as he was walking down a hallway at school. Suddenly he was possessed by a "warm feeling [that] fell on me out of nowhere," accompanied by a sense that he was all right, "everything is all right." It was the most powerful experience of his life. Russell's mother had died when he was twelve. Since that time he had felt rejected by everyone and surrounded by hostility and failure. The hallway experience literally turned his life around. The magic continued. "Looking back," he recalls, "I can see a whole string of unlikely events that had to occur. I also remember moments when new skills seemed to drop down out of the sky, and I felt as if I had a new eye or had tapped a new compartment of my brain."[7]

As a high school player, he was "easily forgettable." But by luck he was picked as a member of a California All-Star team that toured the Pacific Northwest for a month. During this period basketball became an obsession. He absorbed all he could from the other players. Then, after a week on the tour, he says that "something happened that opened my eyes and chilled my spine." He was sitting on the bench, watching every move made by two star players. Then he shut his eyes and visualized the play, and then reran it in his imagination with himself making the moves. When he got into the game, he made the play as if he had become his teammate—like "stepping into a film and following the signs." During the tour he was "nearly possessed" by basketball. On the long bus trips he would make mental replays of the game just played. "It was effortless; the movies I saw in my head seemed to have their own projector, and whenever I closed my eyes it would run." Later he began to imagine playing defense. He visualized himself as the shadow of the player he was trying to guard, a mirror image who moved instantly, as when performing a dance. When he put this technique to work on the court, it was highly successful. Russell became famous for his amazing defensive moves, including blocked shots, which were "not done" in those days. The defensive techniques "grew out of my own imagination, and so I saw them as my own," Russell relates.[8]

As an NBA professional under Celtic coach Red Auerbach, Russell describes several exceptional moments when the team was "in the zone." "All sorts of odd things happened," he states.

> The game would be in a white heat of competition, and yet somehow I wouldn't feel competitive—which is a miracle in itself. I'd be putting out the maximum effort, straining, coughing up parts of my lungs as we ran, and yet I never felt the pain. The game would move so quickly that every fake, cut, and pass would be surprising, and yet nothing could surprise me. It was almost as if we were playing in slow motion. During those spells I could almost sense how the next play would develop and where the next shot would be taken. Even before the other team brought the ball in bounds, I could feel it so keenly that I'd want to shout to my teammates, "It's coming there!"—except that I knew everything would change if I did. My premonitions would be consistently correct, and I always felt then that I not only knew all the Celtics by heart but also all the opposing players, and that they all knew me. There have been many times in my career when I felt moved or joyful, but these were the moments when I had chills pulsing up and down my spine. . . . On the five or ten occasions when the game ended at that special level, I *literally* did not care who had won. If we lost, I'd still be as free and high as a sky hawk.[9]

In Russell's descriptions we see stunning analogies to the eras of medicine. Era I, mechanical-physical medicine, is akin to the "easily forgettable" first stage of Russell's career when he was a tall, unpolished youth who relied on physical talent alone. Era II, mind-body medicine, is analogous to Russell's second stage in which he began to image and visualize his moves and those of others, playing and replaying them in his imagination before translating them onto the court. Era III, nonlocal medicine, compares with Russell's final period when he transcended his individual, separate self and stepped outside space and time, knowing others as himself, knowing ahead of time what would happen, experiencing the game as a timeless flow, oblivious to winning or losing—all in a state of joy.

LOVE AND EMPATHY

What allows a person to shift from a self-oriented way of experiencing the world to one in which there are no boundaries between persons? How was Marilyn Ferguson able to sense her father's death at a great distance, as if she were somehow one with him? What permitted Bill Russell to achieve a sense of becoming one with his teammates? I've read hundreds of letters from readers describing nonlocal experiences such as these, always looking for the key that makes this switch possible. I believe the vital ingredient is *love*—a state of caring and compassion that is so deep and genuine that the barriers we erect around the self are transcended.

People often know when a distant loved one is in the midst of an illness or emergency or is facing death. The sense of connection can be so strong that they experience the same physical sensations that are being felt by the distant person—for example, chest pain if the distant individual is having a heart attack or pelvic pain if she is a woman having a baby. These events are almost always experienced in the context of love. Even "the zone" in sports can be understood in terms of love—love of the game that is so deep that winning and losing become irrelevant, love so strong that "opponent" is transcended in favor of oneness with one's competitors.

Related to love is empathy—putting oneself in the place of others, feeling what they feel. Empathy was undervalued in Era I, medicine's mechanical phase. If drugs and surgical procedures work on their own, who needs empathy? Empathy has recently received another blow with the ascendance of "managed care," as if care and caring could be "managed" like a baseball team or an investment portfolio. The premium in these environments is on efficiency and cost cutting, not in nurturing a warm, lifelong relationship with a patient and his or her family. Physicians working in managed care settings frequently say they don't have time to deliver good *mechanical* care, much less caring and compassion. The time constraints they face are serious. Surveys show that the average time that front-line physicians in managed care environments spend with patients is around seven minutes and is decreasing.

But empathy is a two-way street. Patients, as well as physicians, need to do their share to allow empathy to develop—which gives me an opportunity to express a pet peeve. I practiced internal medicine with a large group of physicians in Dallas, Texas, for many years. Patients would gather in the waiting room, awaiting their appointments with their respective doctors. On passing by the waiting room, I would sometimes hear snippets of conversation. I recall hearing one patient ask another, "Which doctor do you use?" Although I realized it was an innocent expression, I bridled at the thought of being *used.* "Which physician do you *see* or *collaborate* with?" would have felt much better. As a physician, I would have difficulty empathizing with someone who was trying to *use* me.

Let's consider what happens when we go to a physician in search of a diagnosis for a particular problem. *Diagnosis* is derived from Greek words meaning "a knowing that exists through or between." A correct diagnosis, in other words, requires a uniting of doctor and patient so that knowing can flow between them and yield a solution. In order for this process to take place, the doctor must be capable of empathy—the willingness to become someone else and to experience the other person's feelings. This means that in order to be a good physician, a doctor must be willing to become nonlocal, to transcend her personal boundaries. If the physician is too hesitant, proper, professional, guarded—too *local*—to become her patient, she will likely be an incompetent diagnostician.

True, the need for nonlocal bonding between a physician and patient can sometimes be skirted by resorting to laboratory tests to make a diagnosis. But without entering into one's patient, the physician is frequently in the dark about which tests to order. And when all the tests are in, the doctor must still ask, What does this data *mean?* Diagnosis does not come out of raw numbers. It comes about when nonlocal mind comes into play, when the physician and patient become one. To be a great diagnostician, therefore, the physician must actually *enter* a patient—a process that gives new meaning to the question "Is the doctor 'in'?"

If it were otherwise, computers could diagnose as skillfully as physicians. So far, they cannot. Computers are decidedly local objects. They cannot let down their boundaries and transcend space, time, and person to become one with a patient. Computers are "used"; they can be "on" but not "in."

Neuropsychiatrist and neuroscientist Mona Lisa Schulz, in her book *Awakening Intuition,* gives an example of the role of empathy in making diagnoses nonlocally, which she calls "doing a reading."

> One of the first times I did a reading, I imagined myself inside a person's body, looking around. I put myself right in the abdominal cavity, by the aorta, and looked up. The aorta looked like a tree to me. It looked massive. The arteries to the abdomen, the pelvis, and so on were its branches and roots, jutting out at right angles. I got so close to it in my mind's eye that I could actually see the texture of the aorta's "bark."
>
> Now when I do a reading of a person, I first see myself standing in front of a person, checking the individual's head, eyes, ears, nose, and throat. Then I step inside, into the esophagus, and head south. I go for a ride, traveling through the various organ systems and visually

examining their condition. This is a form of empathy, imagining yourself in another person's shoes. If you have empathy, if your heart is open to receiving information, you'll find you receive more of it. Jonas Salk described his own intuitive process similarly. When doing his research, he said he would imagine himself as an immune system and try to reconstruct what it would be like to engage with a virus or cancer cell. This extreme form of empathy allowed him to acquire new insights and design his experiments accordingly.[10]

If Era III thinking makes diagnosis more accurate, it can also make it more complex. According to Era I and Era II concepts, when a physician encounters a patient with certain symptoms, he or she assumes that the symptoms belong totally to that patient and not someone else. But if physical symptoms and sensations can be shared or transferred between distant individuals, as nonlocal mind predicts, how can the doctor be sure of their origin?

University of Virginia psychiatrist Ian Stevenson has reviewed 160 so-called telesomatic cases, in which symptoms are shared between distant individuals. Here are some of the symptoms experienced by the receivers:

- faintness when his brother was dying
- weeping and restlessness when his wife was dying
- committed suicide when his son was dying
- foreboding of evil when her daughter was dying
- sadness, weeping, trembling when his wife was in accident
- an attack of jaundice when his twin brother was dying
- feeling of horror when his son was dying
- feeling of an electric shock when his wife lost money
- horror and fright when her aunt was in danger of fire
- felt a severe blow on head when his sister experienced a head injury [11]

Sometimes clusters of people can receive the same symptoms, spawning small epidemics. In 6 percent of the cases, two or more people picked up on the symptom being communicated, such as two daughters far apart from each other who simultaneously experienced fright at the moment their mother was dying.

INTERNISTS AND ETERNISTS

When I first reviewed the list of symptoms that people experienced in these 160 cases, they reminded me of complaints I'd heard from hundreds of patients early in my career as an internist—complaints for which I could find no cause. At that time I was not an Era III thinker, and I was not familiar with nonlocally communicated symptoms. Some of the cases are still vivid in my memory. Looking back, I strongly suspect they may have had a nonlocal origin. But I was stuck in Era I and Era II. After eliminating physical causes, I almost always settled on a psychosomatic, Era II–based explanation. If I had been able to make the leap into Era III and think nonlocally, I could have been a better physician to my patients.

Neuropsychiatrist Schulz describes how her own physical problem—being hit by a truck while jogging—was communicated nonlocally to her mother. At the moment she was getting ready to cross the bridge in Oregon where the accident occurred, her mother and father were attending a meeting of the historical society in their hometown on the East Coast. Suddenly, in the middle of the meeting, her mother stood up and said, "Ed, something's just happened to Mona Lisa." Her outburst was recorded in the minutes of the meeting by the historical society's secretary, including the time, which corresponded with the moment the truck struck her daughter three thousand miles away.[12]

Psychiatrist Ian Stevenson says, "It seems quite likely to me that many instances of symptoms induced by extrasensory processes are overlooked because this possibility is simply not envisaged." The key, a colleague of mine said, is to think like an "eternist" instead of an internist. An internist always looks internally, inside the patient, for the origin of the problem, in the present moment. An eternist looks at all of time and space, both inside and outside the patient, for solutions.

Putting this approach to work would build better relations between patients and their doctors. Many physicians consider hypochondriacal patients the bane of their existence, and they communicate their disdain. These individuals consume considerable amounts of the doctor's time (and the doctor consumes considerable amounts of their money), to no avail. In the end both parties are usually unhappy with the results—a frustrating string of normal tests, no diagnosis, and a large medical bill. If a telesomatic, nonlocal origin were considered for these types of problems, this might justify a wait-and-see strategy without embarking immediately on an expensive diagnostic workup. Other physicians would prefer to regard telesomatic disease as a "rule out" diagnosis that one considers only when all other reasonable, treatable possibilities are eliminated. In either case, the physician might discuss the situation openly with her patient, who could participate in deciding how aggressive to be with diagnostic measures. Caution is wise, because telesomatic illness, even if present, can coexist with diseases originating solely within the patient.

Again, patients need to do their part in putting this approach to work. Currently, patients often consider physicians who order lots of tests to be more thorough and "scientific" than a doctor who does not. If such a patient comes away from an appointment without tests, he or she is often disappointed. This is typical Era I thinking—the body is a machine and must be taken in from time to time for tests and a tune-up before it breaks down. There is nothing wrong with an Era I approach if used wisely; I'm all for the judicious use of annual exams, mammograms, and blood tests. The problem comes, however, when a patient expects a brain scan when she has a tension headache, or when a physician routinely orders a colonoscopy to evaluate a tummy ache. This sort of overkill is not as prevalent as it recently was—since the advent of HMOs, which reward physicians for *not* doing tests—but it still goes on.

Physicians who make the transition to Era III should hang onto their hats, because this can be a bumpy ride. Doctors can experience their patients' symptoms nonlocally, and this can be unpleasant.

Psychiatrist Schulz describes an example. She was doing a "reading" for a woman over the telephone, entering the woman's body intuitively in an attempt to "see" what was wrong. Although she could discover nothing abnormal, she began to grow increasingly uncomfortable, feeling hot and flushed. She took off her sweater, got up, and adjusted the thermostat in the room, noting that it was already set at a normal temperature. She got back on the phone with the woman and said, "I must be coming down with a cold, because I feel as though I have an incredible fever." The moment she spoke these words she realized that the woman herself was suffering from fever. "As soon as she confirmed it, the heat I was feeling dissipated," Schulz relates.[13]

Schulz does not believe this is a rare, exotic ability. "Whether we call it a telesomatic event or empathy," she states, "an emotional umbilical cord does exist between people, and we pick up messages from one another. Some people," she adds, "are able to become more skilled at this than others, and they call this ability medical intuition. Some people have a somewhat greater innate capacity for intuition than others, in the same way that some people have innate musical ability while others do not. But most people," she concludes, "have the ability to acquire more skill at anything they set their minds to learning. This is as true of intuition as it is of playing the piano."[14]

But we must be cautious. Excessive empathy can cripple a physician. I once knew a pediatric oncologist who identified so deeply with his young cancer patients that he began to attend their funerals and mourn with their parents. He became so depressed he had to withdraw from his medical practice and seek professional help.

Many physicians are aware of the dangers of excessive empathy and respond by walling themselves off from their patients. Yet too little empathy also creates problems. It is perceived by patients as coldness and arrogance, creates barriers to communication, and blocks an avenue for arriving at diagnoses, as we've seen.

I view empathy and the sharing of symptoms with another individual like pain. The ability to sense pain is a wonderful gift that helps us survive. When we touch a hot stove, the discomfort is a valuable source of information. There are diseases that dull one's sensitivity to pain, and those who suffer from them lose fingers and hands because they can't detect when they are in danger. But too much pain can also be devastating and in extreme cases can lead to suicide.

In the same way, we can be either too sensitive or insensitive to the pain of others. Helping young physicians achieve "empathic balance" is a task for medical schools in Era III.

DREAMS

Dreams, we say, are private experiences. Unless we voluntarily share our dreams with others or talk in our sleep, no one can penetrate our dreaming life and know our nocturnal thoughts. Yet dreams are not always the private fortress we imagine them to be. Nonlocal mind is unlimited; it is not restricted to our waking, daytime existence, and it extends to sleep and dreams. Just as nonlocal mind draws us together during our waking moments, it unites us while we sleep. That is why the field of human relationships includes dreams.

CREATIVITY AND PROBLEM SOLVING

Nonlocal connections with others during dreams can pay surprising dividends. For example, problem solving and creative insights often flow back and forth between individuals during dreams. They seem to take place, as we've already seen, between people who have empathic and loving relations with each other.

In her book *Writers Dreaming,* writer and dream researcher Naomi Epel interviewed more than twenty acclaimed American writers about the role dreaming played in their creative process. Novelist Leonard Michaels revealed how he and a close friend influenced each other nonlocally through shared dreams:

> This is going to sound a little mystical but I had a woman friend who, in response to problems that appeared in my life, problems that had

only to do with me, would dream solutions or illuminations of the problem. That is to say she would formulate the problem in her own dreams without my telling her what the problems were.

I was writing something—I had been working on it very hard—and it had to do with popular Latin music, salsa. I hadn't discussed it with her or anybody else except the magazine editor I was writing the piece for. In the midst of this essay, when it was really tough and it had really become problematic, we had dinner and she said, "I had this dream and it was full of musical notes." She started describing the notes and it was the stuff I was writing about! Now where did she get that? She'd never had a dream like that in her life. She doesn't even know music.

I suppose it's quite possible that dreams are expressive of the deepest levels of intimacy between people. At least that has seemed to be the case in my life. When I say between people I mean not only between a man and a woman but even between you and yourself. You, at one point in your life, and you, at a much later point in your life, when you're really a considerably different person. Then you are in an earlier incarnation dreaming your future, your fate.[15]

THE DREAM HELPER CEREMONY

Robert L. Van de Castle, the former director of the Sleep and Dream Laboratory at the University of Virginia and professor in the Department of Behavioral Medicine, has developed methods of purposefully using nonlocal mind, through dreams, to help others solve problems.

Van de Castle's interest in shared dreams began when he participated as a subject in a series of dream experiments at Maimonides Hospital in Brooklyn during the 1960s. The goal of these studies, which were conducted by researchers Stanley Krippner and Montague Ullman for more than a decade, was to determine whether an individual, while dreaming, could receive specific information from someone else. Volunteers in a sleep laboratory would be asked to dream about a picture that was going to be randomly selected after they had gone to bed, a picture that would be focused on by a distant "sender." Or the individual would try to dream about a picture postcard that would not be selected until the following day. The dreamers were awakened when their brain waves and eye movements indicated they were dreaming. Independent judges later decided if there were correlations between the image that was sent and the dreams. Stunning similarities were often seen. In one experiment, Henri Rousseau's painting *Repast of the Lion,* in which a lion is biting into the body of a smaller animal, was selected as the dream target. The dreamer had several dreams about violence and animals. In one dream, about dogs, "the two of them had been fighting before. You could kind of see their jaws were open and you could see their teeth. . . . It's almost as though blood could be dripping from their teeth." For this particular dreamer, the judges confirmed that five of eight dreams corresponded to the image that was sent. The odds against a chance explanation for this outcome were over one thousand to one. The Maimonides studies are classics in dream research, and they strongly suggest that dreams are an avenue of nonlocal communication between separate, distant persons.[16]

"There are more ways of communicating with each other than those acknowledged by current science. . . . [We are all] midnight swimmers in a common cosmic sea," researcher Van de Castle concluded from his participation in the Maimonides dream experiments.[17] Along with Henry Reed, a former psychologist at Princeton University, Van de Castle devised the "dream helper ceremony" so that people can help one another through dreams.

Van de Castle and Reed wanted to provide each individual with an opportunity to learn and grow as a result of participating in the ritual. They discouraged frivolity, emphasizing instead a sense of reverence for the power of nonlocal mind. The strategy was to use telepathic dreaming in a group context to be of service to someone.

In the dream helper ceremony, rather than focusing on a target picture, as in the Maimonides experiments, the "dream helpers" focus on a target person. This individual acknowledges that he or she is troubled about some problem but does not discuss it or give any hint whatsoever as to its nature. At night, before retiring, the dream helpers gather around the designated individual and engage in some activity to create a feeling of closeness and bonding—meditating, singing, silently holding hands, or praying together. The target individual may loan some personal object such as a photograph or piece of clothing or jewelry that enables the dream helpers to form a sense of closeness. That night, the dream helpers renounce their right to experience personal dreams and devote the total activity of their unconscious dream life to the individual in need. They ask that they be used as vehicles for healing and understanding. They may record the dreams they experience that night so as to provide the target individual with every piece of information they've gained. The following morning the dream helpers gather and discuss in detail their dreams from the previous night. "A fascinating pattern emerges as the warp of one dreamer's images is laid against the woof of another's, and dream strand after dream strand is woven into the rich collective tapestry," Van de Castle states.

In conducting dream helper ceremonies on many occasions, Reed and Van de Castle were impressed with the collective accuracy of the dream helpers in identifying the problem for which help was being sought and often coming up with a potential solution.

In one ceremony, black-and-white themes prevailed among the dream helpers' dreams. One person reported driving a black car into the town of White Hall. Other dreams dealt with someone hesitant to accept an Oreo cookie; ordering an ice cream cone with one scoop of chocolate and one of vanilla; the black and white keys of a piano; Martin Luther King, Jr., preaching in front of the White House. Several dreams also dealt with family conflict, dissension, and parental lectures about obedience. The target person, a white woman, was surprised by these dreams because she knew none of the dreamers, and none of them was aware that she was dating a black man and struggling with the question of how to deal with the negative reactions that were certain to come from her family. One dream helper dreamed that his watch was slow, and another dreamed about a movie in slow motion. In their discussions, the dream helpers suggested that the target person proceed slowly and bring up the issue with her family only after making sure of her wish to continue the relationship.

In another dream helper ceremony, after all the dreams had been reported the target person revealed she needed insight about entering a new but undetermined vocation. In almost every instance the dream helpers reported a violent theme in their dreams—wild animals, someone hit on the head with a hammer, and other acts of aggression. Some of the violent dreams dealt with mother-daughter relationships. In one there was a mother duck and several drowned ducklings. When Van de Castle asked the target person why she thought so much violence appeared in the dreams and why they concerned mothers and daughters, she broke down and revealed that her mother, a former psychiatric patient, had been violent and cruel to her as a child. Her mother had tried to drown her once in a tub of boiling water, which might have related to the drowned baby ducks. In their discussion, the dream helpers suggested that the target person consider resolving her longstanding conflict with her mother with the aid of a therapist before moving on to a new occupation.

Skeptics often say that dreams are so general they can apply to anyone's situation and can be interpreted in an infinite number of ways. Van de Castle and Reed do not find this to be the case. The specificity of dreams was demonstrated in a weekend workshop, with Reed working with one group and Van de Castle the other. Although the target person in each group was a female of about the same age, education, and socioeconomic status, the dreams in the two groups diverged markedly. Dreams for person "A" were right on target and did not apply to person "B," and dreams for person "B" were specific for her. Van de Castle says, "It seemed as if each target person was a psychic magnet attracting only dream filings of a very specific metal."

Stanley Krippner facilitates a dream group that meets monthly in Berkeley, California. Krippner was the director of the Maimonides Medical Center's Dream Laboratory and is currently the director of graduate studies at Saybrook Institute in San Francisco. Krippner emphasizes how different his methods are from the way dreams have been handled traditionally in psychoanalysis, which assumes that the analyst understands the dream's symbols better than the patient, whose "defenses" prevent her or him from properly interpreting them. In contrast, Krippner's "group dreamworking" method takes the power away from the therapist and places it in the hands of the dreamer. In the group situation, the dreamer can share as much or as little of what she has learned in the interpretation process as she chooses and can stop the process at any time. The group's function is only to stimulate and support the dreamer in her task of understanding, never to dictate.[18]

Dream helper ceremonies bind individuals together in the common cause of helping someone in need. The dream helpers give freely of themselves during dreams, holding nothing back. A feeling of love, caring, and empathy envelopes everyone concerned—all hallmarks of Era III therapies. Seldom does any single dreamer grasp the full extent of the target individual's problem. But when all the different insights are combined, a solution is often forthcoming.

Dream ceremonies should never be undertaken lightly and never used as mere entertainment. The target person's problem should not be trivial but instead be worthy of the time, attention, and energy of the dreamers. The dream helpers should not participate unless they are willing to commit totally to dealing with the problem.

SELECTION AND TRAINING OF DREAMERS

Superstars and protégés exist in every field, from athletics to mathematics to music. They also exist where nonlocal mind is concerned. In Era III, selection procedures would be developed to identify these highly talented individuals. There is nothing elitist in this. As parapsychology researcher and author Dean Radin explains, we use this rationale in many areas, often for the common good, such as in the extensive selection procedures used with jet fighter pilots. We do not randomly select people off the street and expect them to be able to fly fighter jets. We want to start with individuals who have natural abilities, which can then be honed to a finer edge with training.[19]

Physicist Russell Targ and healer Jane Katra, in their book *Miracles of Mind: Exploring Nonlocal Consciousness and Spiritual Healing,* describe the qualities we would look for in selecting individuals to participate in Era III training programs, and the factors that such programs would seek to enhance.

> Rapport [is] . . . paramount. . . . Commonality of purpose and mutual trust are essential prerequisites. . . . Such agreement and coherence among individuals is often difficult to achieve and maintain. However, in a situation where no personal gain and no specific outcome is sought, a meeting of minds, or resonance, may easily occur. Jazz musicians playing together, with the mutual understanding that the music produced is an ad-libbed group creation, often experience such psychic cohesion. Group meditation or prayer for the purpose of healing are other examples where the merging of individuals' consciousness is easily attained, because there is no consideration of personal profit involved. . . .
>
> Liberation from self-consciousness can also be attained whenever people surrender their individual identities and join their minds together, focusing their attention on creating a common goal, or being of service to others. The trust and rapport . . . can then be quickly achieved.[20]

Targ and Katra deserve our attention. Targ was one of the first physicists to explore "remote viewing" at Stanford Research Institute and later at the CIA, and Katra is a healer with deep experience.

A NATIONAL DREAM TEAM

In Era III, a high-priority task would be to identify people with dreaming talent, who would be asked to dream collectively for the nation. This National Dream Team would function like the above dream helpers who gathered to dream solutions to the problems of single individuals, only the scope of their task would be larger—social, economic, and foreign policy questions. They might be recruited much like professional athletes, with scholarships or other inducements. Training programs as well as examination and certification policies could be developed. The dreamers might function like the oracles of ancient Greece but with much greater precision as a result of their training and cross-comparisons of their dreams with other data sources.

People, through dreams, have often gained information that could have affected the course of a nation and of history. President Lincoln dreamed his own assassination two weeks before it happened. A few nights before his father was shot in 1972, George Wallace, Jr., dreamed the event. In 1914 Bishop Joseph Lanyi of Grosswardein in Hungary dreamed in detail the assassination of Archduke Franz Ferdinand of Austria during the latter's trip to Sarajevo. The bishop had once served the archduke as a tutor and knew him well. If his dream had been taken seriously, the archduke's trip might have been canceled, or he might have been advised not to ride in an open motorcar; or security might have been tightened—and the twelve million military and twenty million civilian deaths of World War I might have been averted.

If a British dream team is developed, one of its members might be Barbara Garwell, a homemaker in Hull, England, who seems particularly adept at assassination dreams. In 1981 she dreamed that a well-known actor was shot by two SS men with a pistol as he got out of his limousine. Three weeks later, John Hinckley, who had once been a member of a neo-Nazi group, shot the former actor and then-president Ronald Reagan. Although her dream is ambiguous, it might have provided enough suspicion to prevent the assassination attempt if taken seriously. In the same year, 1981, Garwell had a vivid dream of men sitting in a single row of seats in a stadium. The men had "coffee-colored skin," wore dark suits, and were situated in the Middle East. Suddenly two soldiers rushed the row of men and sprayed them with automatic-rifle fire. Three weeks later, President Anwar Sadat of Egypt was assassinated by four men in a setting not unlike her dream. Again, her dream, though not exact, contained enough detail that a tragedy might have been averted. In both instances, Garwell related her dreams to others before the fateful events occurred, and witnesses signed statements confirming the sequence of her dreams and the later happenings.[21]

Political leaders and heads of state *already* employ nonlocal methods of gaining information. They often say they arrive at solutions to problems through prayer, meditation, and solitude. A National Dream Team would merely formalize these efforts and bring refinement and discipline to them. This does not mean we would make national decisions solely on the basis of someone's dream, as ancient emperors sometimes did in response to a single oracular vision. In any decision, multiple data sources would be relied on; dreams would be simply one source of information among many.

PRIOR EXPERIMENTS

We have already experimented with something similar to a National Dream Team. From 1973 through 1989, a program on Anomalous Mental Phenomena was carried out by SRI International (formerly the Stanford Research Institute) and continued at Science Applications International Corporation (SAIC) from 1992 through 1994. In this series of experiments, individuals tried to acquire information from a target and to intuit future events, and researchers tried to determine whether these procedures were useful for government purposes. Could they do it? In a report commissioned by the CIA at the request of Congress, Jessica Utts, Ph.D., of the Division of Statistics, University of California, Davis, concluded,

> Using the standards applied to any other area of science, it is concluded that psychic functioning has been well established. The statistical results of the studies examined are far beyond what is expected by chance. Arguments that these results could be due to methodological flaws in the experiments are soundly refuted. Effects of similar magnitude to those found in government-sponsored research at

> SRI and SAIC have been replicated at a number of laboratories across the world. Such consistency cannot be readily explained by claims of flaws or fraud.
>
> The magnitude of psychic functioning exhibited appears to be in the range between what social scientists call a small and medium effect. That means that it is realizable enough to be replicated in properly conducted experiments, with sufficient trials to achieve the long-run statistical results needed for replicability.
>
> . . . Precognition, in which the answer is known to no one until a future time, appears to work quite well. Recent experiments suggest that if there is a psychic sense then it works much like our other five senses, by detecting change. Given that physicists are currently grappling with an understanding of time, it may be that a psychic sense exists that scans the future for major change, much as our eyes scan the environment for visual change or our ears allow us to respond to sudden changes in sound.
>
> It is recommended that future experiments focus on understanding how this phenomenon works, and on how to make it as useful as possible. There is little benefit to continuing experiments designed to offer proof, since there is little more to be offered to anyone who does not accept the current collection of data.[22]

It became quite clear in these experiments that some people were more talented than others. Some individuals were able to describe distant sites in cameralike detail. These are the persons who would be identified for participation in a future Dream Team.

Hal Puthoff, of the Institute for Advanced Studies at Austin, is a highly respected physicist who was also involved with earlier CIA-initiated work in remote viewing at the Stanford Research Institute. In summarizing these developments he agrees with Utts, saying, "Over the years the back-and-forth criticism of protocols, refinement of methods, and successful replication of this type of remote viewing in independent laboratories . . . has yielded considerable scientific evidence for the reality of the phenomenon. Adding to the strength of these results was the discovery that a growing number of individuals could be found to demonstrate high-quality remote viewing, often to their own surprise . . ."[23]

Former President Carter, in a speech to college students in Atlanta in September 1995, revealed that during his administration a plane went down in Zaire, and a meticulous sweep of the African terrain by American spy satellites failed to locate any sign of the wreckage. Without his knowledge, Admiral Stansfield Turner, head of the CIA, turned for assistance to a woman reputed to have psychic powers. As related by Carter, "she gave some latitude and longitude figures. We focused our satellite cameras on that point and the plane was there." Independently, Turner himself also acknowledged the agency's use of a remote viewer.[24]

Describing the hundreds of remote viewing experiments carried out at SRI from 1972 to 1986, physicist Targ states, "We learned that the accuracy and reliability of remote viewing was not in any way affected by distance, size, or electromagnetic shielding. . . ."[25]

It is only natural that government leaders would be concerned with the impact of nonlocal mind on national security. But if remote viewing is not in any way affected by distance or size, as Targ states, its use extends far beyond national security. There may be no limits to its use; the infinitely large and the infinitely small, the past as well as the future, all may be fair targets. As parapsychology researcher Dean Radin says in his book *The Conscious Universe,*

> We might imagine a future "Clairvoyant Space Corps" tasked with exploring distant galaxies. Likewise . . . we may envision teams of Indiana Jones–like "time historians" who explore ancient and future civilizations. We also imagine that mind-matter interaction effects may someday be used to push atoms around, operate psychic garage-door openers, and operate wheelchairs.[26]

What do the skeptics say? Ray Hyman, Ph.D., of the Department of Psychology, University of Oregon, who rejects psychic functioning in general, has offered a dissenting opinion to those above on the subject of remote viewing. But in spite of his objections he acknowledges, "The case for psychic functioning seems better than it ever has been." He admits, "I do not have a ready explanation for these observed effects. Inexplicable statistical departures from chance, however, are a far cry from compelling evidence for anomalous cognition."[27] This latter comment seems to me a strange position to take if someone respects the methods of science. It suggests we are free to disregard data arbitrarily, accepting those findings that please us and tossing out those that don't.

Some of the objections to nonlocal knowing are amusing. In 1974, the CIA challenged Puthoff's remote viewing group to provide data on a Soviet site "of ongoing operational significance" in Semipalatinsk, USSR. The remote viewer selected to "see" the site provided a description that was startlingly accurate. (These results, now declassified, can be judged by anyone.) When Puthoff and his colleagues forwarded the results for analysis, one scientific group declined to get involved because, although the results might be real, the phenomenon was possibly demonic.[28]

Critics often say that remote viewers succeed only in identifying simple targets about which they can make general statements, which give the appearance of success. Not so, states physicist Targ: "We discovered that the more exciting or demanding the task, the more likely we were to be successful."[29]

THE ARTS

I grew up being somebody else—an identical twin. My brother and I were virtually indistinguishable to others and often to our parents as well. Once when we were two years old, I became ill with a soaring fever. We lived on a farm far from town, and home remedies were standard. When they didn't work, my parents, fearing something serious, undertook the long drive to the doctor, planning to leave my healthy twin brother at my grandparents' farmhouse on the way. When they finally arrived at the doctor's office, they discovered that the fever had disappeared—because they had switched babies. They had my healthy brother in tow, and I had been left with my grandparents.

This was only one of many similar incidents, and all identical twins can identify with them. Identical twins learn automatically what it's like to look and feel like another person. I am sure this is one reason why empathy, putting oneself in the place of another, holds such great fascination for me.

But by the time I'd finished my medical training, I'd learned that if an individual feels he's another person, something pathological is going on; he's schizophrenic, a "multiple," or possessed. But we become other people every night in our dreams. Novelist and philosopher Arthur Koestler pointed out that one of the most remarkable features of dreaming is the capacity of the dreamer to be more than one person.[30] In our dreams the boundaries of the self are fluid and blurred. We play multiple roles—what Koestler called "impersonation"—such as watching an execution one moment and being the victim the next. But "impersonation" is not limited to dreams. When a caring parent disciplines her child, she puts herself in the child's place to see how the discipline will be received and what effect it will have. When a surgeon advises a mastectomy, she (hopefully) puts herself in the patient's position and tries to see things from her point of view. We function as doubles many times every day without realizing it.

Becoming another is particularly important in the performing arts. When an actor creates for the audience the impression that he is King Lear, he is assuming a double identity. Truly great performers have the ability to make us feel they are *really* the character they are portraying. When they are successful we feel as if we, too, are the character, which is why our heart races and our pulse pounds during an exciting drama and why a tragedy makes us feel morose.

We leave a great performance feeling drained, inspired, sad, or joyful because we have had a nonlocal experience of becoming someone else. We do not *decide* to become the actor on the stage or screen; the process is unconscious, what Koestler calls an "underground game." We know rationally that we are local, isolated creatures and cannot possibly be the actor, but the unconscious mind knows otherwise. We bumped into this process above in discussing empathy. "Empathy is a nicely sober, noncommittal term for designating the rather mysterious processes which enable one to transcend his boundaries," Koestler says, "to step out of his skin as it were, and put himself into the place of another."

Since the time of ancient shamans, therapists have devised ways of breaking the hold of unhealthy behaviors and habits on the self. This is often done by helping an individual see herself as others see her. Great healers, I'm convinced, do this intuitively. These methods have become formalized in modern psychology, such as in therapies that involve role playing, in which a patient is encouraged to take on the character and feelings of another person, like an actor onstage. If you want to understand why your mother always criticized you unfairly, put yourself in her place—not merely "acting" her but *becoming* her. Allow yourself to behave and speak as she did. Enter her fully without monitoring or censoring your feelings. You may see that your mother, like you, was trapped in behaviors that were extremely difficult to break. You may discover that she behaved toward you the same way her mother behaved toward her. This insight may make her behavior understandable and perhaps forgivable, and it may help heal the unpleasant relationship between you and her. Role playing can be transformative and healing. It is an example of how medicine is related to the arts and why it is an art.

Great therapists have often glimpsed the existence of nonlocal *mental* connections with their patients. "Some eminent psychiatrists—among them Charcot, Freud, Jung, and Theodor Reik—have expressed, or hinted at their belief that not only empathy, but something akin to telepathy operates between doctor and patient in the hothouse atmosphere of the analytical session," Koestler notes.[31]

The greatest healers, I believe, are the greatest actors—not in the sense of faking behaviors and feelings when they encounter a patient, but in the sense of *becoming* the patient, just as an actor becomes the character she plays. Again, this is merely another way of describing empathy—entering another person and feeling what that person feels.

We have learned to use artistic performances as therapy without realizing it. When we feel depressed or out of sorts, we often take in a movie, play, or concert and feel better afterward. But we sometimes feel *worse* afterward; it depends on the nature and the intensity of the experiences of the characters we merged with. Just so, some therapists make patients feel worse. Therefore it is not enough to *become* another; impersonation has side effects, and we must be careful with whom we merge.

THE ERA III ACTORS' STUDIO

When actors assume a role, most of them do so on a make-believe basis. They know they are "playing" a role, wearing a mask. Nonlocal mind, in contrast, is not make-believe; it is a built-in feature of the world, including the world of the stage and screen. At some level, an actor *is* the character she plays. Can we imagine an Era III actors' studio in which actors explore their nonlocal connections with others? In which they allow themselves to *be* another person instead of merely *playing* the role?

A step in this direction has been taken by Turtle Studios, a not-for-profit educational and cultural arts organization in Alexandria, Virginia.[32] Several times a year Turtle Studios sponsors intensive "Spielraum" sessions for actors and artists. Spielraum is a method of combined play and creative work that allows the participants "to rapidly recognize and experience the artistic and creative spirituality that dwells within a person, leading to an appreciation and creation of beauty." During an intensive training period, six to twelve people and two coaches come together for three days. The participants and particularly the coaches often report intense emotional reactions when they are empathically attuned and in sync with one another. They frequently describe these experiences as a "gathering and focusing of energy."

Researcher William D. Rowe performed an experiment to detect whether the focused group energy the actors and coaches report is real. He stationed nearby a field-deployable random number generator (FieldRNG) from the Princeton Engineering Anomalies Research (PEAR) Laboratory. We met a machine like this, a random event generator (REG) in the previous chapter; like the REG, the FieldRNG spits out equal numbers of ones and zeros while running on its own. As in the case of the REG, extensive experiments document that its output can be deflected in one or the other direction by the specific mental intentions of distant individuals or by the focused emotions of groups. Researcher Rowe allowed the electronic device to run before, during, and after various sessions at Turtle Studios. The two coaches were asked to indicate when they felt the group's energy was focused, and their reports were compared with the performance of the FieldRNG. Eleven formal experiments were conducted. Eight of them showed that when the coaches felt that the group members came together as one and focused their energy, the FieldRNG's output indeed deviated from random, becoming more orderly and less chaotic.[33]

It was as if the electronic instrument were functioning as an "empathy detector," responding to the heightened emotional experiences of the actors and coaches when they felt they had achieved a unitive, harmonious state with one another. These feelings appeared to send healing effects into the world, causing the chaotic behavior of the machine to become more orderly, which is a criterion of healing. The coaches indicated that during these moments the actors' performances improved. These findings do not stand alone; they have been replicated by the Princeton group and by others.[34]

These experiments suggest that empathy can be taught and that it changes the world.

Actors routinely use video cameras to tape their performances and refine how they *appear* on stage. The FieldRNG allows them to correlate how they perform with how they *feel.* The FieldRNG is essentially an empathy training device. It shows actors that their emotions reach beyond the stage and that their performances need have no limits—that when Shakespeare said that "all the world's a stage," he was not kidding.

MORE THAN AN ACT

There is evidence that emotions, even though an "act," are associated with medically significant changes in the body. Researcher Paul Ekman, of the University of California, San Francisco, has shown that people who willfully contort their facial muscles into expressions of happiness, fear, or anger will soon come to *feel* happiness, fear, or anger.[35]

Immunologist Nicholas Hall and his colleagues have reported immune changes associated with acting. Their findings suggest that acting is more than an act; it exerts effects on the body that may have important consequences for our health. They measured changes in the immune systems of two actors before, during, and after they performed in two plays. One was a madcap comedy, *Lucy Does a TV Commercial.* The other was a serious play whose main tone was depression, Peter Barnes's *It's Cold, Wanderer, It's Cold,* which is set in turn-of-the-century Russia during the days of the revolution. It takes place in a prison cell on the eve of the execution of an assassin. The researchers started drawing blood and measuring the heart rate of the two performers before they even received the scripts. They monitored the actors through rehearsals and during all subsequent performances. The plays were presented at the same time of day, before a different live audience, daily for two weeks. Hall and his colleagues measured the responsiveness of immune cells called T- and B-lymphocytes. The data suggested a correlation between the type of personality being performed and immune responsiveness. After the *Lucy* comedy, the female performer showed increases in her immune functions. After she performed a depressing role in the drama, these measures were diminished. The male performer projected an anxious personality in the comedy as well as in the serious play. His immune function decreased following both performances.[36]

If we identify with actors who are performing anxious or depressing roles, will we experience decreases in our immune function? Are comedians healthier than tragedians? No one has examined these questions as far as I know, and Hall's study suggests that they be taken seriously. Although we cannot choose *whether* we are nonlocally united with others—nonlocality is a fact, not a choice—we can nonetheless exercise the *degree* to which we participate in many nonlocal experiences, just as actors choose how deeply they enter into roles they play.

If Shakespeare was correct in asserting that all the world is a stage, then we're all actors and every life is an act. The question becomes, which role shall we play? Physician-author Bernie S. Siegel cites a conversation with anthropologist Ashley Montagu, in which he (Siegel) asked Montagu how he could be a more loving person. Montagu replied, "Act as if you are loving."[37] I was surprised when I read Montagu's response, because it sounded like a recommendation for false behavior. Montagu's writings have inspired me for years, and this comment did not ring true with the lofty vision of human potential that runs through his work. Now I believe I see what Montagu meant. For him, acting was more than an act; it was an exercise in being. This implies that we must choose our roles carefully and exercise caution with whom we identify. As writer Marguerite Yourcenar says in her book *Memories of Hadrian,* "The mask, given time, comes to be the face itself."

Era III will witness a fusion of medicine and the arts for one primary reason: *Nonlocal events have local consequences.* Nonlocal experiences leave their tracks in the body. Their effects are real. The arts are more than entertainment; they affect the body, and that is why they cannot be divorced from medicine.

Art is derived from the Latin *artis,* meaning "to join" or "fit together." The essence of nonlocality is the joining and fitting together of things that appear separate. Nonlocality is art and art, nonlocality.

NONHUMAN RELATIONSHIPS

Animals may enjoy advantages over humans when it comes to communicating nonlocally. We humans are obsessed with the idea of a self that is confined to our physical brain and body. This screens out a lot of nonlocal experiences, because we know ahead of time they are impossible. Since animals presumably don't experience a local self and they don't intellectualize, it is possible that they don't experience the prohibitions against nonlocal mind that we do.

Perhaps one reason we are drawn to animals is that we sense in them that the nonlocal form of consciousness is still in flower. Perhaps we need the presence of animals because we need to recover our nonlocal mind, and we sense that they somehow can teach us how.

WHAT IS THE HUMAN-ANIMAL BOND?

When we develop a special relationship with a pet, we experience something mysterious, something not easily understood—the "human-animal bond," researchers call it. What is this bond? What is being bonded? Why does the bondedness feel good, and why is it good for health?

Pets provide us the opportunity to unite unconditionally with another living being. They teach us to love. Love, in a general sense, involves a relaxing of personal boundaries and a willingness to "become one" with someone else. Love, to flourish, requires surrendering our rigid sense of individuality, which creates distance and separateness.

My hunch is that we do not "develop" a bond with animals or with one another; our bondedness is fundamental, natural, factory installed, the way things are. Of course it doesn't feel that way most of the time, because through the processes of individuation and socialization we learn to devalue and ignore our bondedness with others. Pets bring us back to the realization of our unity with other living things and help us remember who we are. One way they do this, I believe, is through their uncanny ability to encourage us to remove our masks and behave naturally. As Samuel Butler (1835–1902) said, "The great pleasure of a dog is that you may make a fool of yourself with him and not only will he not scold you, but he will make a fool of himself too."

A SPIRITUAL CONNECTION?

"There is in every animal's eye," said English reformer John Ruskin, "a dim image and a gleam of humanity, a flash of strange light, through which their life looks at and up to our great mystery of command over them, and claims the fellowship of the creature, if not of the soul."[38] Ruskin implies a spiritual connection between animals and humans, and some modern scientists agree with him.

Aaron H. Katcher, a physician at the University of Pennsylvania School of Veterinary Medicine, suggests that pets perhaps catalyze a spiritual component to healing. He and his colleagues found that 98 percent of dog owners spent time talking to their dogs, 75 percent thought their dogs were sensitive to their moods and feelings, and 28 percent even confided in their dogs. Katcher believes people derive benefits from these interactions not unlike those from prayer. "Without being irreverent," he states, "it is possible to think about the similarities of the comforts of prayer and the comforts of talking to an animal. Prayer is frequently accompanied by sensual enrichment such as incense, music, special body postures, the touch of folded hands or rosary beads, just as dialogue with an animal is accompanied by the enrichment of touch, warmth, and odor. In both instances," he concludes, "the talk is felt to be 'understood.'"[39]

An example is reported by Kathleen MacInnis, a primary nurse on the cardiovascular unit of Northern Michigan Hospitals in Petoskey, Michigan. MacInnis relates her experience in caring for Dorothy, a young girl who was dying. Dorothy's blood pressure was sustained by intravenous medication, and her pain was managed with frequent morphine injections. What could the nursing staff do to make Dorothy's final hours more joyful? They decided to bring her beloved dog, Blackie, a black Labrador, up to the floor for a visit. Dorothy smiled at the plans but warned that Blackie was an eighteen-month-old "wild and crazy teenager" who would tangle everything up, sniffing all around. When Blackie ventured into the coronary care unit, his nails clicking on the tiles, he seemed fascinated by all the new smells. As he and a nurse walked down the corridor, Blackie's healing magic came alive as people smiled at him and reached out to stroke his fur and pet him. As they entered Dorothy's room, he took one crook-eared look at his mistress, gave one happy sniff, and climbed in bed beside her—burrowing in and snuffling his way from hip to armpit as he had learned to do previously. Dorothy was too sick to scratch his ears as she usually did. But she asked the nurses to do so, because she knew what Blackie wanted. "As we scratched his ears," MacInnis says, "his eyes rolled partway back into his head in ecstasy; this pleased Dorothy the most." Later that week Dorothy died in her sleep—following coma, a balloon pump, a ventilator, and her visit with Blackie. "We felt good about that," MacInnis stated, "but wondered if we had waited too long for her to enjoy the visit, to scratch his long black furry ears. So much—too much—of what medical science does for people isn't really *good* for them. At least," she continues, "it doesn't bring them happiness. Doesn't really comfort them." She adds, "It seems to come down to a matter of semantics: the difference between a dying myocardium and a broken, lonely heart." She concludes, "I reflect on all the things I cannot change or help my patients with. But I will remember with pride the evening Blackie spent with Dorothy. And how, in the face of incurable disease, her broken heart was healed."[40]

A case demonstrating the ability of a pet to create a clinical turnaround was that of T. S., a thirty-nine-year-old full-time writer and part-time English professor, who was admitted to the critical care unit with fever, generalized muscle weakness, malaise, and tenderness and swelling in both lower extremities. The problems arose after he had begun a strenuous physical training program a week earlier, including weightlifting and jogging, through which he was determined to become "lean and mean." He was suffering from acute renal failure caused by excessive physical training, rhabdomyolysis (dissolving of muscle tissue), and dehydration. Shortly after being admitted, hemodialysis (kidney dialysis) was begun.

T. S. did not tolerate his hospital experience well. Following admission, he became emotionally upset and withdrawn. He had never been sick before and wondered if he would survive. As his feelings of desperation deepened, his mood proved contagious, for his wife began to take on his despair. In an effort to help him find hope and motivation, his nurses began to think creatively about how to change things. They discovered that one of his primary sources of joy was Mike, his three-year-old pug dog. When questioned, T. S. revealed deep concerns for his pet. Mike had been a constant companion, including during his recent jogging program. What would happen to Mike? Would he, too, get sick because they were running together? T. S.'s nurses decided it was crucial that Mike come for a visit, which was agreed to by the nursing and medical directors of the treatment unit. Before Mike could enter, the infectious disease team was consulted to develop guidelines.

When Mike saw his master, he became so animated he almost wriggled out of the arms of T. S.'s wife. The two friends visited and napped together. One by one, the members of the nursing staff found an excuse to come by the room to visit and chat. T. S. became visibly more relaxed and energized. The visit was so therapeutic that the staff allowed it to become an every-other-day event. Thirty-six days following admission, T. S. was discharged to return home.

In today's high-tech health-care environment, it is ironic that a puppy's sloppy kiss can create measurable health benefits. Yet the evidence favoring the health value of pets is so compelling that if pet therapy were a pill, we would not be able to manufacture it fast enough. It would be available in every hospital, clinic, and nursing home in the land. When a patient entered such a facility, the opportunity to have contact with a caring animal would be routine. In view of the data that is available, we can well ask, What's the holdup?

We hear on every hand that medicine is in decline and that our system of health care is going to the dogs. May it come to pass. Literally.

CHAPTER 4

ERA III IN EVERYDAY HEALING

After ecstasy, the laundry.

ZEN SAYING

How can we apply the lessons of Era III in our lives? The first step involves learning to trust nonlocal mind, the principle on which Era III medicine is based.

As an example, consider the tragic problem of sudden infant death syndrome (SIDS), in which an infant dies in his or her sleep for no obvious reason. In 1993, researchers at the Southwest SIDS Research Institute in Lake Jackson, Texas, performed a survey of parents who had experienced the death of a baby from SIDS. The survey also involved parents of normal infants. Both sets of parents were asked if they had ever "sensed" that something was going to happen to their baby. Twenty-one percent of the SIDS parents and only 2.1 percent of the control parents reported that they'd indeed had a premonition that something was going to happen to their infant. To make sure their findings were on firm ground, the researchers asked another control group of two hundred consecutive patients in a suburban pediatric clinic if they had ever sensed that their baby was going to die and then did not die. Only 3.5 percent replied yes.

The SIDS parents said that when they reported their premonition of their babies' impending death to a medical professional, spouse, or friend, they were rarely taken seriously. Despite their requests to their physicians for additional medical evaluation and intervention, nothing beyond routine measures was advised for *any* of the SIDS infants. When asked to take special precautions for the apparently healthy child, the physicians' responses ranged from "outrage" to "placating denial."[1]

This is a sad example of how the failure to trust nonlocal mind can lead to disaster. If the parents had had sufficient trust in their premonitions, they would not have taken no for an answer. They would not have let themselves be pushed around by their physicians and would have found another doctor. But the parents can hardly be faulted. The main problem lay with the physicians. They were trained in an Era I mode, which regards seeing into the future as impossible. Locked into Era I, they considered premonitions offensive and reacted against them with a visceral response. They behaved in the worst possible way—venting their biases and hostilities on those they were supposed to serve. Unable to think outside Era I, these physicians compounded the parents' grief by being unable to discuss their premonitions after their child's death. This added a needless burden to the parents, who were tormenting themselves with whether their child might still be alive if only they had been more demanding or sought a second opinion.

If the physicians had been sensitive to the principles of Era III, they might have taken the parents' premonitions seriously and some of the SIDS deaths might have been prevented. I am not suggesting that all the worries of all parents are valid, but neither are they uniformly false. The challenge for a physician is to develop the sensitivity to tell genuine premonitions from the invalid ones. Physicians may insist that medical decision making should be based on hard evidence, not premonitions. But in fact only about 15 percent of medical interventions are supported by solid scientific evidence.[2] The rest of the time we physicians rely on subjective reasons—including hunches and premonitions—in making diagnoses, deciding which tests should be ordered, and selecting treatments.

It is too easy, however, to blame the doctors. The researchers found that the premonitions of the parents were rejected also by spouses and friends. You don't have to be a professional to be locked into Era I. The prison of Era I can confine anyone.

The challenge for patients and parents is to learn to trust their nonlocal sources of knowing and to refine them. Unless they *are* refined, one can easily fall prey to an overactive imagination. The *reliability* of information is key, whether that information comes through local or nonlocal avenues. And reliability can be fostered by mindfulness—paying attention to the endless instances in which nonlocal mind crops up in our everyday existence.

When the researchers asked the SIDS parents about their experience, the parents revealed that eventually they had come to trust their instincts, in spite of having been rejected by their doctors, spouses, and friends. This is how most people learn to apply nonlocal mind in their life. Through trial and error we sort out valid and invalid pieces of information that come to us nonlocally in the form of hunches, intuitions, premonitions, and dreams. Learning to apply nonlocal mind, therefore, is surprisingly simple. One doesn't have to attend seminars and workshops to develop this ability; one isn't required to become fluent in the language of science or to become an expert in the latest experiments. One merely has to be attentive to the flow of life's events. Each time we have a premonition, a hunch, or a dream, we note it and compare it with the outcome. Our life becomes our lab, and each experience becomes what scientists call a "single case report." Many people find that keeping a written record of these experiences sharpens their awareness and refines their sensitivity to the nonlocal operations of the mind. Like a singer, we need practice in order to learn to sing on key instead of being out of tune when using our nonlocal potential.

Now let's look at additional examples in which nonlocal mind springs to life in everyday experiences. I have assigned the events to certain categories, but these divisions are artificial. There are no absolute separations between premonition, insight, intuition, dreams, hunches, synchronicities, revelations, telepathy, and clairvoyance. The common denominator is nonlocal mind—a way of knowing that operates outside of space and time.

TELEPATHY

Telepathy is a term coined in 1882 by the British researcher F. W. H. Myers to describe communication between minds by some means other than the normal sensory channels. Telepathy is an expression of nonlocal mind—mind manifesting beyond the confines of brain and body.

Telepathic, nonlocal connections often spring vividly to life when a loved one is sick. Terry Partington, of Billings, Montana, experienced the following when his son Ben fell ill:

> Upon arriving at a new job assignment in Montana, I called my wife in Texas, where my family temporarily remained, to tell her that I had arrived safely. I was shocked to learn that my fourteen-year-old son, Ben, had been admitted to the ICU earlier in the day due to high blood sugar, later diagnosed as diabetes. I immediately returned to Texas, and, when the situation eventually stabilized, I returned to Montana to my work.
>
> Later that month, my wife dropped Ben off at a video store around 8 P.M. and told him to be home by 9. When he had not returned by 10, she began to worry and went looking for him. She returned empty-handed and called me, and I assured her everything would be fine.
>
> I could not get back to sleep and therefore stayed up for a while. Suddenly a blinding and highly emotional thought entered my mind that I should tell Ben to wake up. I started repeating the command over and over and could not think of anything else. Soon I was in tears with no apparent explanation. This went on and off for at least an hour. I then became sleepy and went back to bed.
>
> About 3 A.M. my wife called to say that Ben had just returned and was groggy but otherwise okay. He had been walking home in a forest near our house and suddenly felt ill, sat down on a log to rest,

and passed out due to what later was determined to be low blood sugar from too much insulin. After some unknown period of time, he woke up, immediately took a glucose tablet he always kept with him, and returned home.

Some would say that my action and Ben's awakening are distinct and unrelated events. I have a deep feeling inside of me, an assurance, that this is not the case.[3]

DREAMS OF ILLNESS

Intuitions of illness often occur in dreams. However, we don't dream of neon signs announcing "You Have Cancer." The messages are usually veiled in metaphors and images. Neuropsychiatrist and neuroscientist Mona Lisa Schulz, in her book *Awakening Intuition,* reports several dreams that were prophetic of health problems. The images in these dreams illustrate the language of the dream world.[4] One man, who was a heavy smoker, dreamed he was back in the army. Under attack, he took cover in the hollow of a large tree. Enemy bullets cut through the tree and penetrated the lower left side of his chest. He later found he had a small tumor of the left lower lobe of his lung, where the bullet had hit him.[5] In another case, a woman dreamed she was lying on the ground when the earth began to give way beneath her and formed a cavity, in which she began to suffocate. It was discovered two months later that she had cavities in her lungs from tuberculosis, which by then was causing her breathing difficulties.[6] In another case, a woman dreamed repeatedly, for almost a year, that a nurse was holding a lighted candle to the dreamer's leg, as if trying to show her something. Eventually the woman began to feel pain in her left shin; afterward a bone infection was discovered at the site, requiring surgery.[7]

This woman's dream image of a nurse carrying a light is legendary. It is also reminiscent of Florence Nightingale (1820–1910), the founder of modern secular nursing, who revolutionized hospital care, introduced the use of statistics in health care, and promoted public health approaches in preventing disease. Nightingale carried a lantern on her hospital rounds while ministering to British soldiers in Turkey during the Crimean War (1854–1856), as described in Barbara Montgomery Dossey's illustrated biography, *Florence Nightingale: Mystic, Visionary, Healer.* The enduring image of "the lady with the lamp" was immortalized in a poem about Nightingale by Henry Wadsworth Longfellow.[8]

Dreams sometimes tell us to change certain health habits before illness actually develops. William Dement, who did pioneering work in sleep and dreams, had been a two-pack-a-day smoker. One night in a dream he saw an X ray of his own chest. The entire right lung was cancerous. A colleague in the dream examined him and confirmed that the cancer had spread to his lymph nodes. Dement was deeply anguished during the dream. He knew his life was over and that he would never see his children grow up. On waking, he experienced inexpressible joy to discover that it was only a dream. He felt reborn to life and immediately stopped smoking.[9]

Some argue that dreams are not nonlocal events but are merely "brain noise," the nocturnal recall of memories that are scrambled into meaningless gibberish. But there's more to it than that. "Brain noise" theories cannot explain how a healthy individual could know about impending health problems that lie months or years in the future.

Some illness-related dreams are Era II events in which the mind and body are in close communication. An example is a man who dreamed about a rat gnawing at the right upper side of his abdomen. During the day he experienced indigestion, but this got better with changing his diet. A week later he had the dream again of the rat gnawing at his stomach, and the next day he noticed deep pain and tenderness at the site of the rat's attack. He visited his physician, and tests revealed an ulcer at the specific location.[10]

INTERMEDIARY OBJECTS IN HEALING

The possibility that physical objects might act as a go-between in conveying healing intentions is an ancient idea. For millennia shamans and folk healers have blessed talismans and charms of various sorts before giving them to the sick person, in the belief that they are able to mediate the healer's wishes.

We modern physicians have our own unique talismans, which may be conveying our healing intentions without our realizing it. The intermediary objects take the form of white coats, stethoscopes, imposing gadgets, and colorful, oddly shaped pills. These objects are drenched with symbolic meaning for patients. In true Era I form, we deny that these physical objects could take on our healing intentions and bring about healing when our patients are exposed to them, but evidence suggests otherwise. As we saw earlier, in a series of controlled laboratory experiments, Dr. Bernard Grad of McGill University demonstrated that the intentions of a healer could affect tumor growth and wound healing in animals and that the same effect could be generated if the healer blessed a piece of cotton wool and placed it in the animals' cages without ever going near them.[11] Somehow the physical object seemed able to mediate healing in the healer's absence.

The ability of physical objects to convey healing intent is another way in which Era III medicine can be translated into our lives. Consider the experience of Lisa, the teenaged stepdaughter of Meredith Goodman of Columbus, Ohio.[12] In July 1991 Lisa was involved in an automobile accident that fractured her neck at the C2–3 level with injury to her spinal cord. The neurologist caring for her was so convinced she would die that he withdrew orders for an MRI. He placed her in tongs, put her on a respirator, inserted a feeding tube, and sent her to the intensive care unit so her family could have one final visit before she died.

The family was told Lisa would probably not survive the night. She could not swallow, could not breathe on her own, and had no feeling or movement from the neck down. One physician explained that Lisa had only a thin thread of spinal cord not completely severed; she should not even have survived the trip to the hospital and the emergency treatment. The doctor asked the family to pray for a miracle, but, he added, "I've never seen a miracle."

Lisa did not die as predicted, so a tracheostomy was done. Although there was no improvement, she remained alert and tried to talk. The family learned to read lips, but Lisa became angry when she was unable to get her words across. She hated the ventilator, and her one wish was to get off it. Goodman describes what happened:

> There is an old tradition among some religions of a prayer cloth. A group of Christian people, usually a church or congregation, pass around a piece of cloth. Each participant holds it in their hands and says a prayer of healing. The cloth is then given to the person whose healing is being prayed for. It is to be placed on the area of the body where healing is desired. In Lisa's case my father's church anointed a prayer cloth for her. Dad brought it to the hospital and tied it around Lisa's upper right arm.
>
> She would not allow this cloth to be removed. It became soiled with spilled tube feedings and medications. Still Lisa refused to have it removed. She refused even to let it be washed out because she was afraid the prayers would also be washed out in the process.
>
> Lisa asked Dad to pray especially for her breathing. Once again a prayer cloth was anointed and prayed over, this time for her breathing to return. At the same time, her maternal grandmother's church also anointed a prayer cloth. Dad's second cloth was tied around Lisa's left arm and her grandmother's was pinned to her chest.
>
> Lisa's healing did not occur overnight. In fact, four years later it is still progressing. Today Lisa breathes without a respirator, eats a normal diet, does her own transfers, balances with a quad cane, and, once upright, can take a few steps with a little help with her balance. She uses a computer and operates her electric wheelchair. She is learning to drive a specially equipped van and attends college alone. Her progress amazes everyone who hears her story. Her neurosurgeon can only shake his head in disbelief over each new development. I think he would now say, he has indeed seen a miracle.
>
> I am amazed how Lisa does not see herself as handicapped—limited, yes, but not in any way abnormal. From the first day after her accident Lisa has had a calm acceptance of the changes in her life. Her faith is what I see as the true miracle. I pray for faith like she has.

Another case involving an intermediary object involved eighteen-year-old Tim, who was hospitalized with severe chest injuries following an auto accident.[13] His pulmonary status deteriorated, and he developed acute respiratory distress syndrome (ARDS). Tim was eventually intubated and placed on a respirator. Unable to talk, he wrote the name *Willie* for his mother. She knew immediately this meant Willie Nelson, Tim's favorite country-and-western singer. His mom immediately brought his Willie Nelson tape collection and a cassette player to his bedside, along with two red bandannas, Willie Nelson's trademark. Tim had his mom tie one bandanna around his head and another across the palm of his right hand. He was deeply attached to the objects and would not allow them to be removed, even when his clinical condition began to worsen. Just before he died he raised his hand with the red bandanna and gave his dad a "high five," a gesture that had special meaning between them. His parents realized that this was Tim's signal that everything was okay and that they should let him go.

The two examples illustrate different ways of employing intermediary objects—deliberately trying to infuse them with special power, as with the prayer cloth, or allowing the object to function symbolically, as with the red bandannas. These cases are not rare. Anyone working in intensive care units knows that families are always bringing objects to the bedside that have special significance for both patients and their relatives—stuffed animals, photos, crucifixes, and so on. From the Era I perspective, this practice is a worthless nuisance. From the Era III viewpoint, it is to be encouraged, and families should be asked to deliberately imbue the objects with intentions for healing, like Lisa's prayer cloths.

DREAMS OF HEALING

Intimations of healing often occur in dreams. A typical healing dream took place for thirty-six-year-old Ken Engler of Kenosha, Wisconsin, when conventional treatment had failed:

In June of 1993 my neurologist found a tumor in the right temporal lobe of my brain—a benign but fairly aggressive astrocytoma. In July I had my first brain surgery, followed in August by six weeks of radiation therapy. I recovered well except for seizures due to residual scar tissue from the surgical procedures. In January 1994 I had my second surgery to the right temporal lobe to remove the scar tissue. This time recovery was much slower. In May 1994 I was diagnosed with an inoperable and terminal brain tumor—inoperable because this time it was in an area of the brain that was inaccessible without causing severe permanent damage, and terminal because no other treatments were available. I had already received a maximum dosage of radiation, and chemotherapy held no hope. My neurosurgeon told me I should get my personal affairs in order as soon as possible.

Shortly thereafter I began using water from Lourdes. I had developed a pipeline for this holy water directly from the spring in France. I would massage the water to my scalp and whisper a prayer to Mary, our Mother of Lourdes, asking for her intervention in my illness. I did this once or twice a day, every day. I truly believed this was holy water that had the power to heal. Thirty days later my next MRI scan showed the tumor had begun to shrink. I continued to use the holy water, and the shrinkage continued.

Prior to my last MRI in November 1995, I had a very unusual dream. It was about a woman named Mary (mother of Jesus?). The woman was plainly dressed and about my age (36). The unusual aspect of this dream was the incredible love I felt from this woman. It was the most unusual feeling I have ever had. It was not a sexual love but one that resembles a mother's love for her child. I did not believe it was possible to feel such deep love and warmth. The last thing I remember Mary saying to me before I awoke was, "Ken, be happy." I awoke from this dream with a comfort and total absence of anxiety I had not felt in two years of illness. This dream and the sensations associated with it lingered for several days. I could not think of anything else.

> I went into the MRI chamber on November 23, 1995, with Mary's love still warm in my heart. Later that day the results showed the tumor had virtually disappeared![14]

Elaine Haire, of Las Vegas, Nevada, also had a dream visit from such a healer, which may have saved her life:

> I had been chronically depressed through most of my life. Depression ran in the family; a younger brother committed suicide, and I had made a couple of suicidal gestures at ages twelve and thirty. I was in Jungian analysis and struggling desperately, in torment. At night I would write symbols on my body and scream, "I can't change if I can't see!" Within a short period of time I had a dream/experience of an Indian medicine man performing a ritual over me that included chanting, dancing, and shaking a rattle. I didn't remember the words, but when I woke I knew I'd never have to deal with that degree of psychic pain again. I'm no longer chronically depressed, and my life is increasingly filled with joy.[15]

INTUITION AND REVELATION

Joseph Campbell, the great mythologist, once said that we gain wisdom by one of two ways. We may have a *revelation* and become wiser suddenly, or—far more common—we can *suffer* and gain wisdom gradually. When insights arise suddenly and unbidden—Campbell's first method of acquiring wisdom—they often do so through dreams. These can be so profound they seem to originate nonlocally from outside the self, from an infinitely wise source. The following is an example related by Harold McKinney, of Miami, Florida:

> I was dreaming, and I knew that I was dreaming. Nothing that I can remember was happening in the dream, so I began to pray that Christ would come. As I prayed, a pond formed in front of me. From this pond came a goose covered with down. (My daughter has Down's syndrome and was about one year old when I had the

> dream.) The goose came and sat in my lap. It was warm and felt so good to hug. I really enjoyed hugging the goose, but I soon realized that this goose was very large and that I couldn't move or get up with this large burden in my lap. While wondering what I could do next, I looked down and realized, though the goose wasn't flapping its wings, we were both in the air. Needless to say, at this point I no longer considered the goose to be a burden but grabbed the goose around the neck and held on for dear life. I knew that if we were to fall, only one of us knew how to fly, and it wasn't me.
>
> I awoke from the dream, and the goose has been teaching me how to fly ever since. Little did I realize before the dream and prayer that I was the goose's burden rather than she mine.[16]

Skeptics often say, "If nonlocal mind is real, why don't people get rich? If they can know the future, why don't they win the lottery?" Irene Spencer, of Hamburg, New York, did just that. She dreamed of the winning numbers in the lottery—not once but twice. In February 1996 she wrote,

> My life is (and has been) full of miracles. . . . I dreamed of a four-digit number and played the Win-4 game in the New York State Lottery, won $2,600, and paid all my bills. That was some years ago. I dreamed another number and have been playing it for about three years, and it hasn't come in yet. However, since the number is 1225 (which is, of course, the day we celebrate Christmas), maybe the dream means I should be betting my life on Jesus.

Spencer reported again in October 1997:

> One of these days I expect I will be able to tell you about winning money betting on #1225. And, yes, I do *too* pray that it hits. When it wins, it will be both prayer and dreaming!

She proved to be a prophet. Her note of April 1998 says,

> #1225 came in on the New York State Lottery Win-4. I'm paying bills now, and I'm going to get a new roof on my house.[17]

Isabel Allende, the acclaimed South American novelist, has a healthy respect for nonlocal revelations. She relates,

> Dreaming of snakes in Chile means money. I have had that dream twice and I'm not a gambler. I hate gambling. The first time, all the family, all the tribe, was spending some vacation time on the beach in Chile before the military coup, and I had this dream that nineteen snakes were crawling up my brother-in-law's legs. So the next day I said, "You should go to the casino and gamble because you will get a lot of money playing nineteen." And he said, "No, I'm terrible at that." I said, "Well let's split whatever it is, whatever you lose or gain." So we went to the casino. I bought a comic book and I sat outside to read while he went in. After a while he came out and he said, "I lost everything. Your dream was the shits." And I said, "But did you play nineteen?" And he said, "No. You didn't say that." I said, "I did say that. There were nineteen snakes!" And so we went back together and he played nineteen and he won! We were so surprised and so appalled at the same time that we started yelling and screaming. We left all the money there and it came up a nineteen again. So we had a basket full of bills, which of course with the inflation is nothing, but at that time we could invite the whole tribe, thirteen people, for dinner in the best restaurant in town. And then I won three hundred bucks in a casino in Aruba with my stepfather after I dreamt that snakes were crawling up his body also.[18]

ANTICIPATION OF DANGER

Nonlocal knowing can manifest during any form of physical danger, such as when a crime is being committed against one's person or property. The following experience is that of Bonnie Johansson, a travel writer in Sydney, Australia:

> Several years ago I was asked to participate in an experiment by a physicist, which required me to keep a dream diary for a year. Each morning I would write out my dream, then phone two independent

people to tell them about the dream, and then see what happened. I found out to my amazement that I was regularly having precognitive dreams [dreams of future events] of such detail and specificity that accident or coincidence seemed to be ruled out.

An example was a dream I had one Tuesday night. I dreamed we were in southern California and had gone out for the night, when two men broke in to rob our house. They took all of my clothes out of my cupboards and drawers and threw them into the corridor so that the corridor was an ankle-deep sea of colour. They didn't take anything, and they didn't damage anything. They got in through the back door, and the back railing was dripping with blood. In the dream when we got home we called the police, and this exceptionally charming and good-looking police captain arrived. I was struck by how charming he was and kept saying to Peter, my husband, "Isn't he charming." He was dark-haired with a moustache. I [wrote down] the dream, rang my two independent observers, one of whom is a psychoanalyst friend, and we tried to decipher the psychological significance of it in my life at that time, but it just didn't make sense.

On the following Friday, Peter and I were supposed to go to the movies. The unannounced short film turned out to be a travelogue on southern California. About 9:00 P.M., I suddenly felt the impulse to tell Peter about my dream, which I hadn't mentioned before.

This turned out to be about the time they actually were breaking into our house. We got back from the movies to find the corridor an ankle-deep sea of colour from all the clothes they had emptied into it. Peter said, "But how could they have got in?" and I said that in my dream it was from the back. As I said it, I suddenly remembered the blood dripping from the back railing and got worried. So we went to the back, and that was indeed where they had forced their way in. It had been raining so there was water, not blood, dripping from the railing. They hadn't taken or damaged anything, even though there was a video recorder and an expensive camera sitting in plain sight. It was a weird break-in. Then we called the police, and lo and behold, guess who turned up—the charming dark-haired police

> captain from my dream. It was really a strange feeling; I kept thinking he must know me because I felt as if I knew him from my dream. Of course he showed no sign of recognition! (Obviously he hadn't dreamed about me!)[19]

PROPHECY

Events such as this, in which people see the future in dreams and visions, resemble prophecy. Prophecy—from Latin words meaning "to speak before"—is a nonlocal mental ability that has been recognized throughout human history. The word *prophet* conjures images of bearded, cranky Old Testament males, but this is only one way prophets have been packaged through the ages. They still exist; Bonnie Johansson is only one among countless examples. These individuals never call themselves prophets, and they often go to great lengths to disguise their talents because they fear ridicule or "treatment" by the psychiatric profession. One of the benefits of Era III is the opportunity for them to come out of hiding as the principles of nonlocal mind gain greater acceptance.

One of the most common examples of prophecy in medical settings is when patients see their own deaths, even when medical professionals taking care of them see no reason why they should die. As British physicians A. N. Exton-Smith and M. D. Cantaub report in the prestigious medical journal *Lancet,*

> Seven patients had a premonition of death, and this was communicated to the nurses by such remarks as "Goodbye, I am going" an hour before death. Another thanked the staff nurse who was doing the medicine rounds for all she had done, and said that she would not need tablets anymore after tomorrow. A man with congestive heart failure thanked all the nurses for their attention the day before his death, and a woman with rheumatoid arthritis, half an hour before she died, asked that her friend should be summoned. There is no doubt that these patients became aware that they were about to

> die, but the manner in which this knowledge was imparted to them could not be ascertained.[20]

SYMPATHY

Sometimes we appear to be linked nonlocally not just to humans but to physical objects. Our ancestors called these "sympathetic" connections, because they reveal that we are somehow "in sympathy" or resonance with the object. Such an incident happened to Dr. Connie Hernandez, a naturopathic physician from Mountain View, California:

> When I was twenty-eight years old, my husband and I started trying to conceive and praying for a child. I had previously undergone treatment for a chronic pelvic inflammatory condition and had been advised to have a hysterectomy. Tests showed that my fallopian tubes were blocked. While traveling in Japan and throughout Asia, we became involved in many things, including photographing fertility dolls and visiting local healers. Follow-up tests found my fallopian tubes to be unblocked, but we still did not conceive.
>
> In the summer of 1986 we traveled to Indonesia and found our way to a small Indonesian antique shop on the island of Lombok. In the shop we discovered an amazing fertility doll. She was carved in wood, about two feet tall, pregnant, with human hair and pubic hair and prominent genitalia. Her name was Jujuk, we were told, and she had been carved more than 100 years previously and had sat in a fertility temple of the hill people. The shop owners emphasized her rarity, in that she was completely whole and undamaged. The shop owners gave me permission to photograph her. As I directed the camera at her, I felt in some way bound to her and had to have her . . . a bit ridiculous, since we were traveling with backpacks, had just started our trip, and would have to carry her around with us. Nonetheless we purchased her, wrapped her in newspaper, and set off down the street.

The street was crowded with people, animals, and motorcycles, as Asian streets often are. We had not gone far when a motorcyclist ran over my right foot. There was a moment of pain and a terrible crunching sound, and I found myself catapulted out of my body. I observed as my husband turned toward me, and the motorcyclist came to a screeching halt—a frozen moment in time. Perhaps just seconds later, I reentered my body and was astonished to find that I was in no pain, and was walking fine. We walked back to our hotel room, unwrapped our prize, and discovered that the doll's right foot had been crushed.

Back at the antique shop, it was explained to us that dolls of Jujuk's nature are carved to link with a particular human being. Jujuk had been carved nearly one hundred years before my birth. It was thought most auspicious that Jujuk had already demonstrated her link with me, and I was advised to sleep with her near me. I was also given a specially prepared herbal potion by the grandmother of the antique shop family.

We eventually carried Jujuk with us back to Seattle, where we moved for our naturopathic training. When we were moving from one home to another in 1987, an antique dealer came to the garage sale we were having. He came directly to me, looked me in the eye, and asked me if I had any fertility dolls for sale. Without a moment's hesitation, I walked into the house, brought out Jujuk, and sold her. As the man walked down the drive, I realized I was pregnant. Since I had tried to conceive for nine years, and counted many days of many cycles, it hadn't occurred to me that I might be pregnant.

Gabriel, my son, is now eleven.[21]

Sympathetic magic has been used throughout human history to influence other people. According to this idea, if one wishes to control a distant individual, one has only to manipulate an object belonging to them—a lock of hair, nail clippings, or a piece of their clothing. Because the object and the person were once in contact, it is believed that the object acts as a go-between and carries the influence to the actual person, similar to the use of intermediary objects in healing, as we saw above.

Sympathetic magic is one of the oldest expressions of nonlocal mind—consciousness working outside the confines of the body to affect someone else, with a physical intermediary. We pride ourselves in having rejected the old beliefs, but they are still with us. When a sweetheart clips a ringlet for her lover who is going to war, she is employing a physical object she hopes will keep them connected nonlocally. And what is *sympathetic magic* if not another term for Kodak? When parents carry photos of their children in their wallets, they sense that the photo has the power to bind them to those they love. The blizzard of photos exchanged between distant individuals all over the world is evidence that we still believe, at some level, that physical objects have the power to unite us nonlocally.

TELESOMATIC CONNECTIONS

Experiences in which we share physical sensations with a distant individual, as if two persons are somehow sharing a single body, are extremely common. These have been called "telesomatic" events, from the Greek *tele,* meaning "far off," and *somatikos,* referring to "body."[22]

Linda Whitson, of Mansfield, Texas, a registered nurse working in home health care, experienced a typical telesomatic connection with one of her patients:

> I have had several telesomatic experiences over the past few years. These events have involved my children, close friends, or patients I feel bonded to. All of the occurrences have involved a death or potential life-threatening danger.
>
> For example, I received a call about 1 A.M. regarding a patient I had been seeing on weekly basis. Her daughter had already called 911. I went to my patient's home and had never seen her looking so

> bad. Her blood pressure was quite low, and she was having chest pain and difficulty breathing, which had not responded to the medications given by the emergency medical team. After the ambulance left with her, I returned home to bed and sleep. I was suddenly awakened by a violent jerk that went through my whole body. As I was trying to figure out what happened to me, the phone rang. My patient's daughter told me over the phone that her mother had just arrested but that the doctor was able to "shock her back"—a medical procedure that causes precisely the sort of violent jerk I experienced.
>
> Even though I have always been religious or spiritual, experiences such as this have given me a greater appreciation of the "unknown." I feel I have become more connected to all that is seen and unseen.[23]

Sometimes we know something so deeply—"in our bones" or "in our gut"—that it is accompanied by actual physical symptoms. These are some of the most common telesomatic events. An example took place in the life of Mary B. Boardman, a former hematology technologist of Willseyville, New York:

> One of my strongest telesomatic experiences was the day my father died. I awoke in such a state of extreme agitation that I called in sick to work. I paced all day, aggravated and upset for unknown reasons. A voice told me to "call home," but I ignored it (I don't do that now!). By five o'clock I was so upset that I decided to lie down. As soon as I did, the voice in my head screamed to me to *Call home! Now!* I did. My mother answered the phone in a totally distressed, panic-stricken voice. She said my father had just fallen through her arms onto the floor, and she called the ER team. Then, I knew what it was all about. I knew he was dead. I called my brothers in Seattle and Philadelphia and told them, before knowing from the doctors. They understood the knowing that my father and I always shared, and they made their arrangements. It was an hour later when they called to tell me he had died.

The joke of the family was that he and I didn't need a telephone to communicate. It was true. We still don't.

SYNCHRONICITY

One day, while writing my 1989 book, *Recovering the Soul*,[24] I was sitting at my desk describing the spiritual implications of nonlocal mind. My focus was on distant, intercessory prayer and how prayer is infinite, universal, cosmic. Barbara, my wife, brought to my desk a large bouquet of cosmos, one of our favorite flowers, which she had just cut in the garden. As I admired them, one of the stems started moving—a praying mantis, which my wife had not noticed. I felt a shiver up my spine. One of my themes—prayer in the cosmos—had literally sprung to life. *Mantis* is derived from a Greek word meaning "prophet," and prophecy, knowing the future, was another nonlocal theme I had been exploring. It was as if the praying mantis was a cosmic prophet who had come to say, "You're on the right track. Don't hold back!"[25]

INTUITION

Barbara Mathews is the director of Multisensory Learning of Toronto, Canada, which presents customized seminars and keynote addresses to businesses. The goal is to help participants go beyond intellectual approaches in mastering their work and relations with coworkers. Mathews's approach is based on research showing that if people use all their senses they are more likely to experience breakthroughs in understanding, and their long-term retention of concepts and skills is increased. She uses play-acting, out-of-this-world simulations, visual cues, dialogue, and tactile and kinesthetic exercises to shake up her clients and boost them into new learning grooves. She may, for example, invite a manager to reenact a temper tantrum and express his anger not through words but by gyrating and blowing on a kazoo. People learn to sense and understand their "hot buttons" and express conflict safely. The result is better team dynamics, conflict resolution, and workplace creativity.

Mathews's techniques have attracted a lot of attention, and her business is booming. How has she done it? She dreamed things up—literally. "For six years I've been growing my business largely inspired by dreams and prayer," she says. "Everyone told me, 'Forget it, you're crazy,' but the inner visions just seemed so compelling, real, and immensely rewarding."

Mathews is also an inventor, and she designs the interactive training tools she uses in her seminars. One of her inventions, an interactive mask, had adjustable headgear that was held together by Velcro tabs. This allowed the wearer to adjust the tension of the elastic band for comfort. The problem, however, was that the wearer's hair kept getting stuck in the Velcro. Mathews tried countless alternatives, but nothing seemed to work. The day before she was to take the two hundred mask bases to the sewing factory to have the headgear attached, she still had not solved the problem of the Velcro tabs. Finally at bedtime she said, "Holy whatever, I need an answer in the morning." Then she went to sleep. "I had a dream," she states, "with two hundred beautiful white birds flying gracefully in the sky. When I zoomed in for a close-up I noticed that the birds had changed into two hundred flying white bras." Then she woke up and began to dress. "As I was fastening my bra, I suddenly noticed that the bra slider on the shoulder strap was the answer I was seeking for my headgear. Not only did it solve my original problem, but wearers could also adjust the headgear while wearing the mask." Excited about this new possibility, she quickly took the mask bases to the sewing factory. "The head seamstress examined this new alternative fastening method and approved it. And it saved twenty-five cents per unit!"[26]

Creative breakthroughs in dreams are often facilitated by a sense of urgency, as in Mathews's case. Where do the solutions originate? Was Mathews incubating her answer unconsciously? Perhaps, but dream solutions often appear so whole and fully formed, and so unconnected with our own experiences and memories, that they seem to come from a source of wisdom greater than our own. This suggests that knowledge can be shared nonlocally between distant individuals who may already possess the answer we need; or perhaps that knowledge can be drawn from a timeless pool of information.

MAKING SENSE OF HEALING

We cannot make sense of our lives unless we acknowledge that our mind operates nonlocally. If we accept nonlocal experiences as a natural part of the world, more things fall into place and have meaning.

Illness is one of the experiences that we most want to make sense of. When I began to practice internal medicine fresh out of training, I assumed that I would develop, with time, an increasingly rational understanding of why people get sick and why they get well. It did not turn out that way. For me, the mysteries of healing deepened with each passing year. It was not until I began to explore the dimensions of Era III that the vagaries of healing began to fall into place. I do not claim to understand these conundrums, only that my post–Era I ignorance is less than before. I believe Era III can help anyone make sense of healing, especially those events we call miracle cures. There is every reason to focus on them, for if we understood them, we might then take steps to make them more frequent.

A survey of the medical literature suggests that any illness, no matter how hopeless, can up and disappear.[27] We physicians find these cases baffling and sometimes irritating because they are lessons in humility; they say there is a lot we don't understand and can't control. To be comfortable with miracles, doctors must have a high tolerance for the unpredictable and the irrational, which we disdain.

Era III is the breeding ground of miracles. They are usually mediated through some manifestation of nonlocal mind—through intercessory prayer, for example, or following dreams and visions or a sense of contact with an ineffable "other." One of the features I adore about miracles is that they happen to sinners as well as saints. One doesn't have to wear a halo to have a miracle cure.

Rita Klaus developed her first symptoms of multiple sclerosis (MS) at age twenty, while a nun. Unable to meet the demands of convent life, she was given dispensation from her vows. She finished her college degree in biology and became a science teacher in a junior high school outside Pittsburgh, Pennsylvania. She married and had three children. Her disease progressed rapidly, and she was soon confined to a wheelchair, with complete paralysis of both feet and ankles. Contractures and spasticity caused structural deformities of her legs, along with intractable sciatic pain. Certain she would never walk again, her doctors surgically severed the tendons in her legs that kept her kneecaps in place. This permitted her to hobble short distances inside her house with full-length leg braces and forearm crutches.

Although formerly an extremely religious nun, Klaus's spiritual beliefs evaporated like a snowball on a hot stove. "It's a bunch of malarkey," she said. "God does not intervene in the natural order. Every time I see these televangelist creeps, I feel like puking." Despite her rejection of her former beliefs, she allowed her husband to take her to a church service where healing was performed, during which she was prayed over, touched, and hugged. She felt "the strangest experience." As she described it, "There was just this white light, a feeling of absolute love like I'd never felt coursing through me. I felt forgiven and at peace. I wasn't physically healed but had peace of heart, of knowing I was loved and could weather anything."

The years passed and her disease got worse. Her doctors told her there was no hope of improvement and that her nerves and tissue were damaged irreversibly. But in the ensuing years her faith returned and she resumed her prayers. Then one night she heard a voice extending an invitation to her to be healed. She awoke the next day, went to a class in her wheelchair, and began to feel sensations of heat and itching in her lower legs. To her astonishment she realized she could wiggle her toes. Thinking this must be muscle spasms, she dismissed it. Then, when she got home, she bent down to remove her leg braces and noticed that her right kneecap, which had been in a deformed, sideways position since the tendons were cut, had somehow migrated back to its normal position. She said later, "I just remember screaming, 'My God, my God, my leg is straight!'" Then she shed her remaining braces, removed her socks, tucked up her skirt, and said to herself, "If I am cured, I can run up the stairs!"—which she did, all thirteen steps.

Her next venture was outdoors. She ran from the house through the woods, jumped a creek, and came back covered with leaves and mud. She called her priest, shouting, "I'm healed! I'm healed!" The poor man thought she had gone mad. "I want you to sit down, calm down, take some aspirin, and call your doctor," he stammered. Still babbling, she called a girlfriend, who came over and shared a good cry. The following Monday her husband took her to the rehabilitation hospital. Writers Caryle Hirshberg and Marc Ian Barasch, in their book, *Remarkable Recovery,* describe what happened.

> The doctors who convened to examine her were flabbergasted. As the nurses scurried for her charts and patients gawked and craned, the doctors' reactions tellingly varied: "One of my doctors saw me and started to laugh. He thought I must have had a twin who I'd brought in to play a practical joke on him." Her neurologist, she says, was "so angry! He said there is no cure for MS, no such thing as miracles. He even called people at the hospital and told them I was a fraud and a fake."

There seemed to be little doubt about Klaus's cure. Dr. Donald Meisner examined her and found no trace of MS. When word of the case leaked out, he said to a local newspaper, "Spontaneous remissions of multiple sclerosis are possible. The only thing that doesn't fit here is that usually the permanent damage that had occurred up to the point of remission does not go away. In Rita's case, every evidence I could see would suggest that she is totally back to normal."

In contrast to the emotional discomfort experienced by some of Klaus's physicians, others were happy about her recovery. Her urologist, who had last seen her with her bladder swollen to many times normal size and incontinent, retested her and confirmed that the organ had returned to normal. "He said there was no way he could explain it, that it was the most beautiful thing he'd ever seen in all his years of practice, and then he cried." Hirshberg and Barasch obtained one of Klaus's neurological reports from around that time. It read,

> Totally independent of any equipment . . . She has regained full strength of both lower extremities. . . . Her deep tendon reflexes are all symmetrical and normal. . . . A tremendous recovery. I am not sure where to place it in this short period of time. The patient did not get tired of demonstrating to me how good she was. . . . I am very happy. . . .[28]

The sheer unreasonableness of Klaus's recovery is striking. From what we know about how bodies function, no one should *ever* regain full function and normal reflexes in their legs after their knee tendons have been surgically severed. As writer Kat Duff has said, we need to see "through the filter of scientific materialism that clouds our vision . . . to glimpse the deeper mysteries and hidden designs of illness" that lie in that "underworld . . . [where] impossible events, strange visitations, and unexpected transformations" take place.[29]

Such an unexpected transformation happened to four-year-old Ann O'Neill, who was hospitalized with acute lymphatic leukemia during Easter week in Baltimore in 1952. At that time, this disease was 100 percent fatal. The priest had given her a final blessing, and her aunt had already prepared her burial garment, a hand-stitched gown of lovely yellow silk. When the head pediatrics nurse asked the little girl if she'd like to go to heaven, Ann's feisty mother said, "No, Sister, not yet." Dr. John Healy, a pediatrics resident at the time, has vivid recollections of the all-consuming faith of Ann's mother: "She never even questioned for five seconds that this girl was going to get better." Shortly thereafter, Ann's parents bundled her up and took her out of the hospital, into the rain, to the cemetery where Mother Elizabeth Seton, a revered Catholic nun, lay buried. There, surrounded by praying nuns, they laid her on the tomb and asked for a healing.

Back in the hospital, blood tests a few days later showed no trace of leukemia. The physicians were baffled. When word of the apparent miracle reached Rome, Vatican investigators trooped to Baltimore to see for themselves. Nine years later, the Church insisted that Ann undergo a bone marrow biopsy to confirm her cure. The investigation was overseen by Sidney Farber, the well-known Harvard pathologist who had pioneered the first effective treatment for leukemia. Hirshberg and Barasch describe the subsequent chain of events:

> The Pope declared her a miracle, and not long afterward canonized Mother Seton an American saint. Is there a scientific explanation? Ann's doctor, Milton Sacks, one of the country's foremost hematologists, testified at the Vatican tribunal that given the bloody sores on her neck and back, severe anemia, and 105-degree fever, she should not have survived a disease that was then "inexorably fatal."[30]

Oddly, this case was never written up in the medical literature. Sacks told a reporter in 1993, "The only reason this case has not been written up is that I have been afraid to."[31]

The way unexpected, remarkable, spontaneous, or miraculous cures are dealt with borders on scientific and intellectual dishonesty. Consider, for example, how survival statistics are analyzed. A 1926 British study of 651 women with untreated breast cancer revealed an average survival time of 3.2 years following diagnosis, and a median (the point at which half the patients had died) of 2.3 years. A year later a Boston study found an almost identical average survival of 3.3 years and a median survival of 2.5 years. Five years later, another study revealed a median survival of 3.3 years. But this figure, the latter study mentions in passing, was obtained by "omitting two cases of alleged duration of 40–41 years."[32] Hirshberg and Barasch point out,

> These latter cases are what statisticians call "outliers," routinely omitted from calculation because they fall too far outside the

> boundaries of the statistical playing field. Thus out of a study of sixty-four patients, two exceptional—even revelatory—cases are dismissed with two flicks of a parenthesis.
>
> It is something of an irony. Science in general, and the physical sciences in particular, have achieved their greatest successes through the study of rare natural phenomena. The examination of a few odd radioactive minerals altered history forever; Gregor Mendel's solitary wonderment at how pea plants with albino petals could produce offspring with purple ones—in effect, a spontaneous remission of white flowers—led to the feats of genetic prestidigitation that now astound our age.
>
> If anomalies in the mineral and vegetable kingdoms can so profoundly change our world how much more so exceptions in the human realm?[33]

Muriel Bourne-Mullen, a seventy-one-year-old geriatric nurse, ran into the brick wall of Era I medicine with her miracle cure. She grew up in the India of the nineteen twenties and thirties, the daughter of the British Army officer who arrested Mahatma Gandhi. She was nourished on the spiritual lore of Hinduism and often witnessed tribal religious ceremonies. Although a devout Catholic all her life and devoted to daily prayer, she honored all spiritual faiths. In 1987 she discovered she had liver cancer that had metastasized to her lungs. Her doctors gave her six months to live and told her that her condition was so far advanced that treatment would be "quite unnecessary."

Although Bourne-Mullen had a place in her belief system for the mysterious and the unexplainable, she considered herself a critical thinker. Her childhood in India had exposed her to methods of healing unknown to Western science but that she could not ignore. In addition, her subsequent experience as a cancer ward nurse prepared her for the unexpected and the unpredictable. Following her diagnosis, she began to pray more diligently and passionately, beseeching St. Jude, patron of lost causes. Her family rallied around her, and others began to pray. In a few months she inexplicably began gaining weight. Her tumor shrank, then disappeared, confirmed by scans, X rays, and biopsies. In spite of this, her physician felt the need to drag her confounding case back inside medical boundaries and urged her to undergo a liver transplant. She indignantly refused. "I was perfectly hale and hearty, love. I told him there was now nothing the matter with me and that would be quite unnecessary."[34]

Her case was reported in *Gut,* a journal whose title is a clue to the Era I tenor of its reportage. In the article, Bourne-Mullen is fleetingly described as "a sixty-three-year-old white woman with a four-month history of abdominal discomfort and bloating after eating." Then she disappears, replaced by pleomorphic tumor cells with bizarre giant nuclei. The report confirms the original diagnosis of metastatic liver cancer, explaining that because "no treatment was considered worthwhile, the patient was discharged home with no drugs." Only two published reports in medical history of regression of primary liver tumors exist, the article states, and only one, from China, concerns *metastatic* liver cancer. A litany of possible factors are listed that may have been involved in the regression—endocrine and immunologic mechanisms or interruption of the tumor's blood supply. But in the end the physicians throw up their hands and simply acknowledge, with admirable honesty, that "the patient received no treatment for the tumour, and regression therefore can be truly described as spontaneous."[35] What is *not* admirable, however, is the information embargo on Muriel Bourne-Mullen herself—what *she* thought was responsible for a cure, including her spiritual experiences, beliefs, and prayers.

WHAT PATIENTS THINK ABOUT MIRACLES

What *do* patients believe is important when they experience a remarkable recovery? While researching their book, *Remarkable Recovery,* Hirshberg and Barasch surveyed some four dozen patients who had recovered from cancers that should have been fatal.[36] Most of these patients emphasize the importance of prayer (68 percent), meditation (64 percent), and faith (61 percent) as significant to their recovery. Others stressed the value of exercise (64 percent), guided imagery (59 percent), walking (52 percent), music or singing (50 percent), and stress reduction (50 percent). Belief in a positive outcome was emphasized by 75 percent of the patients. Seventy-one percent believed that a fighting spirit was important, but an equal number (71 percent) felt that acceptance of their disease was crucial. This information clearly indicates that there is an "urge toward nonlocality" that is experienced by those undergoing remarkable recoveries—a desire to step outside the self and transcend the ego.

FALSE HOPE?

One of the perennial complaints of scoffers is that discussing miracles creates false hope. Of course any intervention can create false hope if it is oversold—penicillin, for example, which is worthless against an increasing number of infections for which it was once considered a miracle. Or exercise, which can have serious, even lethal, side effects if pursued unwisely. But this does not mean we should abandon the use of penicillin or stop recommending exercise; neither should we retreat from discussing miracles.

Researchers and authors Jeanne Achterberg and G. Frank Lawlis are two of the great architects of the mind-body medicine of Era II, and their current work is filling in the unknowns of Era III. They have spent a combined total of a half century teaching research design and statistics at medical schools and have served as principle investigators on National Cancer Institute grants. In their view, "Depriving cancer patients of any treatment that meshes with their worldview or belief system and is probably helpful (at worst, benign) is unethical, inhumane, and unkind. . . . [Honoring psychosocial and spiritual factors] will create a vastly different science, a different psychology, and a different medicine."[37] The late Lewis Thomas, who was director of research at Sloan-Kettering Cancer Institute and who was once called "the most listened-to physician in America," also emphasized paying attention to remarkable cures such as those above. Thomas noted,

> Spontaneous remission of cancer patients persists in the annals of medicine, . . . a fascinating mystery, but at the same time a solid basis for hope in the future: if several hundred patients have succeeded in doing this sort of thing, eliminating vast numbers of malignant cells on their own, the possibility that medicine can learn to accomplish the same thing at will is surely within reach of imagining.[38]

Critics often make a last-ditch stand against radical healings by claiming they are not really miracles at all. One day, they say, we will be able to explain them by ordinary physical processes, which will prove they were not extraordinary to begin with. For my part, I hope they are right; I hope we *will* discover the intervening variables and use them at will, as Thomas predicted, so we can engineer these events at will. But if we manage to do so, it is likely that the boundaries of the miraculous will move farther out and farther in, into dimensions where space, time, and causation take on new meanings. Here another kind of reasoning will be needed—one that recognizes the unifying links among all minds, that connects past, present, and future, and that annuls the finality of death: the domain of Era III.

ORDINARY MIRACLES

When we think of miracles, we usually equate them with the disappearance of intractable diseases such as in Rita Klaus's case. But there is a spectrum of miracles, from the sensational to the subtle. Little miracles are common in everyday life—for example, the headache or aching joint that "just goes away," never to return. These little miracles are woven so naturally into the fabric of daily life we don't notice them. Zen Buddhists have an expression that captures the natural alternation of the majestic and the mundane: "After ecstasy, the laundry."

It is the ordinariness of Era III and nonlocal mind that increasingly captures my attention. I am fascinated that a housefly seems able to read my mind and safely take flight the moment I decide to swat him. I am impressed that my computer functions without complaint when

I'm writing articles with upbeat themes but goes haywire when I'm writing about negative ones. Or that a misfiled scientific paper will somehow appear when I need it. At such moments all of life seems blessed and it is difficult to tell the ecstasy from the laundry.

As a final example of how we can bring Era III into our lives, therefore, I want to emphasize ordinariness—how the simple act of paying attention, of just sitting with one's problem, can issue in healing.

Writer Kat Duff battled chronic fatigue and immune deficiency syndrome (CFIDS) for years, with recurring fever, exhaustion, loss of memory, fatigue, and aching muscles. At the height of the illness she was unable to do anything but lie in bed. Her book about her experience, *The Alchemy of Illness,* is a literary gem.

Duff's exploration led her into the domain of Era III, a dimension that defies "the rules of ordinary reality," a world "full of . . . impossible events, strange visitations, and unexpected transformations." Once when she was in India and suffering from severe dysentery, a Hindu holy man dressed in orange appeared in the corner of her hotel room and sent healing waves of energy through her "ragged body." A similar event happened when she was sick with CFIDS. When disabled with fever and exhaustion, she experienced an apparition of her uncle, who communicated love to her at a time when he himself was in a coma.[39]

Rhea A. White, a researcher of "exceptional human experiences," considers Duff's discoveries "a possible world-shaking insight into the heart of healing . . . [that illustrates] the necessity of a state of chaos, of liminality, of being in a nether world where anything is possible, and the unexpected can intervene while the expected is unable to function. Metaphorically, if not physically, illness can shake up a person's molecules, setting them off in a furious dance, and when at last they reassemble, the person is born anew, with new awareness, strengths, sense of purpose, and even a vocation."[40]

Many contemporary writers speak almost glowingly of illness, as if it is a nifty shortcut to "transformation" and "growth." Most of these accounts sound as if they are written by people who have never been sick. Duff's graphic descriptions are a warning against selling illness short. Hers was a wretched experience that she almost did not survive. She was constantly confronting her "pack of black dogs . . . despair, envy, and hate," which sucked her into the depths of depression. She hated life and began to behave hatefully to her friends. She felt "small, dirty, and tainted. . . ." Finally she decided to "stop complaining and just sit with my dirt."[41]

As often happens in the nonlocal world of Era III, dreams proved a key to her survival and recovery. In a dream she saw that faith must be practiced, and she began to pray daily. Around the same time she was given an antidepressant drug, with further improvement. She regarded her own prayer and the physician's prescription as "simultaneous reflections in the mirrors of my body and soul." Perhaps it was a coming together of opposites—the spiritual and the physical—that made her healing possible. "Magic is present in those sacred spaces where opposites touch. . . . Miracles of healing can occur at these intersections, although they are not necessary or inevitable, but simply demonstrations of grace."[42]

After three years of sickness, Duff had a dream that revealed that her illness "had come to serve the function of initiation, it was bequeathing me a mantle of power."[43] For her, as in traditional societies, the initiation was a bridge to maturity and greater power in life. The clan, the tribe, the society, and the entire species depended on the collective transformation of individuals. She emerged from her illness in the same way that an initiate in a primitive tribe emerged—"a different person . . . feeling a sense of resolve around the facts of my life, and a compassion for all the parts and players, that has never gone away."

One of the predominant assumptions of Era I is that disease is the enemy and should be avoided at all costs. In Era III, in contrast, another view unfolds, in which the dark side of life must be taken on. Duff discovered through her dreams that healing involves a kind of acceptance, which is not to be confused with passivity and giving up. Accepting illness involves a "reunion of opposites," as she put it—health *and* illness, pleasure *and* pain, light *and* shadow. Duff came to realize that "to heal ourselves and renew the world, we must . . . touch the untouchable within and without ourselves, and hold it as precious. . . ."[44]

THE HEALING OF THINGS

In our everyday life we can apply Era III principles in healing not only people but inanimate objects as well. The key, it seems, is love.

We often adore inanimate objects as if they are alive. Any handyman knows this. I have a pruning saw with which I've had a love affair for ten years. It fits my hand perfectly, and when I use it I wear it like a glove, as if it were custom-made. When the wooden handle broke five years ago, I lovingly repaired it as if it were a child with a fracture. When it becomes dull, I have it sharpened by a professional. Never would I consider discarding it, because I consider it an extension of myself. Just thinking about it makes my connection with it come alive; I want to stop writing, turn off my computer, and go look for a tree to prune. In other words, I have a nonlocal relationship, a shared identity, with my pruning saw. Anyone who enjoys using fine tools can identify with my experience.

The ability to develop intimate connections with material things is especially apparent in the relationships kids develop with computers. They often treat them as if they are alive. As biologist Lyall Watson observes,

> Children exposed for the first time to computers are very concerned about whether the machines think and feel, and soon enter into social relationships with them, getting competitive, scolding, punishing and rewarding them as they would human playmates. . . . One thirteen-year-old says: "When you program a computer, there is a little piece of your mind, and now it's a little piece of the computer's mind. And now you can see it!" And in the end, as a result of growing up with computers, a significant number learn to love the machine for itself.[45]

Of course not all our feelings toward machines are positive. In a divorce case from California, a woman named an IBM "personal" computer as a codefendant. She felt neglected because her spouse spent all his spare time programming. When she found him kissing the machine goodnight, she vacated the house and the marriage.[46]

The urge to try to heal inanimate things appears innate, as when a child puts a Band-Aid on a doll and commands it to get well. But as we grow older we become imprisoned by Era I thinking. We develop a condescending attitude toward material objects and consider them dead. Yet the things of this world continue to beckon us, particularly when they require healing. Some people are eager to answer the call. When Watson was being interviewed in a radio station in South Africa, a piece of equipment broke down. A producer entered the room and brandished a scruffy white overall at the console, which promptly lit up again. "It works every time!" the man said. "The coat used to belong to a senior service engineer who seemed to be able to restore anything to life just by showing up. Now that he's retired, we find the coat a useful stand-in."[47]

Robert Morris, professor of parapsychology at the University of Edinburgh, also tells of such an individual. When a research colleague of his experienced a breakdown in a cryogenic magnetometer, he called a repairman. When the repairman walked into the room, it started up again. As he walked out, it shut down once more. They brought the man in and out three more times—and each time he got near the instrument it functioned perfectly.[48]

Oskar Malecki's neighbors in Gdansk, Poland, call him "Saint Reliant" because he makes house calls to help them start their cars on cold winter mornings. "My boy has special powers with machines," his mother says. "His touch fixes them as if by magic. . . . He repairs church buses for free." On one occasion Oskar got four cars running in less than two minutes by merely hovering over each, humming like a perfectly running engine.[49]

The opposite of Saint Reliant is the individual who is naturally *un*therapeutic, in whose presence things fall apart—a Doctor Death of machines. "At one company, a system would crash during critical client preparations whenever a certain engineering manager appeared. The problem was solved by excluding him from its room whenever an important demonstration was being prepared," writes historian Edward Tenner in his book *Why Things Bite Back: Technology and the Revenge of Unintended Consequences.*[50]

The sense of shame apparently can be used to heal machines, as it is used in the rehabilitation of people mired in alcohol and drug abuse. Harold Ilford, who lives on the east coast of England, watched his neighbor having trouble with her lawnmower. She tried unsuccessfully a dozen times to get it started, then disappeared into her garage. She emerged with an old rusty push mower, which she put alongside the sleek new machine. "I see you're reverting to traditional methods," Ilford remarked. "No, I'm shaming the new one into starting," she responded, "by making believe that if it won't work I'll use the old one instead." And the new one started up on the next trial.[51]

Threats also seem to work. Michael Shallis, a physicist at Oxford University, tells about a "computer healer" he met in Britain. "When a fault occurs in a machine," says Shallis, "his colleagues go and fetch him. He switches the machine back on and it functions normally. It has been known that the very mention of his name corrects the fault: 'If you don't work we'll get Peter and he'll stick his cold screwdriver down your circuits!'"[52]

Throughout history, the mystics, visionaries, and saints of all the great religions have expressed their belief in the unity of *all* things—living and nonliving, animate and inanimate. Era III resuscitates this ancient realization and puts it to work in everyday life.

A WORD OF CAUTION

Nonlocal mind is not a plaything. It can be used not only to help others but also to harm them, which makes it necessary to consider the ethical implications of Era III.

Currently, more than ten million Americans try to harm others with their thoughts. That's the conclusion of a 1994 Gallup poll, which found that 5 percent of Americans pray for harm for others—and that's just the one in twenty who will admit it.[53]

We can also harm others *un*intentionally. Consider how parents use their mental intentions and prayers for the welfare of their children. They often pray, for example, that their children never get sick. If these wishes and prayers were answered, they would be lethal, because the only way children develop a competent immune system to protect themselves later in life is through repeated infections with bacteria, viruses, and fungi.

During the past two decades, considerable evidence has emerged from scientific studies that we can indeed harm other living things with our mental intentions, wishes, and prayers—at a distance, without their knowledge. These studies do not involve humans for a good reason: It is illegal to perform experiments whose goal is to harm or kill people. Therefore, these studies involve *parts* of humans such as red blood cells and cancer cells; or nonhuman entities such as bacteria, yeast, fungi, germinating seeds, plants, mice, and other animals. I have discussed extensively the scientific evidence for "nonlocal harm" in my book *Be Careful What You Pray For,* so I shall give a brief summary of only a few of the relevant studies here.[54]

- Leonard Laskow, an American gynecologist and healer, was able to inhibit the growth rate of cancer cells, as measured by their uptake of radioactive thymidine, a standard index of DNA synthesis, using appropriate laboratory controls.[55]
- Researchers at New York's Mt. Sinai School of Medicine studied two qigong masters, who claim they can send either positive or negative qi (an ancient Chinese term for energy). In a well-controlled experiment they were able, at a distance, to inhibit the phosphorylation reaction of myosin light chains, a biochemical process occurring in human muscle tissue.[56]
- British healer Matthew Manning was able to exert dramatic negative influences on cervical cancer cells growing in culture, using

appropriate laboratory controls. The negative influences occurred not only when he held a flask containing the cancer cells in his hands, but also when he tried mentally to influence them at a distance.[57]

- Jean Barry, a physician-researcher in Bordeaux, France, in a controlled study, demonstrated that people could inhibit the growth of a destructive fungus, *Rhizoctonia solani,* at a distance of 1.5 meters, using only mental intentions.[58]
- Barry's study was replicated by University of Tennessee researchers William H. Tedder and Melissa L. Monty, using the same organism, at distances of one to fifteen miles, using student volunteers as influencers.[59]
- Carroll B. Nash performed a similar study at St. Joseph's University in Philadelphia using the common bacterium *Escherichia coli,* with dozens of unskilled volunteers.[60]

The fact that these studies involve nonhumans is important, because this means that the results cannot be dismissed as due to suggestion or to the power of negative thinking. These experiments provide compelling evidence, in my judgment, that we have the power mentally to harm and sometimes kill living cells, including human tissue, at a distance.

Some people are horrified by this possibility. But Era III honors these abilities and uses them to *improve* human conditions. Sometimes we *need* to harm one thing in order to help another. For example, when we intend, wish, or pray for someone to recover from cancer, we are asking that the cancer cells be destroyed or rendered inactive. When we ask for someone to be cured from an infection such as pneumonia or AIDS, we are hoping to kill, harm, or neutralize the offending viruses or bacteria. When we intend or pray that someone recover from coronary artery disease, we're hoping to obliterate the obstructions in the coronary arteries that are blocking the flow of blood and oxygen to the heart. In certain situations, therefore, the ability to cause harm is a blessing.

What about *protection* from the negative thoughts of others? In Era III we don't have to hold our breath in fear of the next hex, curse, or negative prayer. If we evolved the power to harm others nonlocally, we almost certainly evolved inborn protective mechanisms as well. These innate forms of protection I've called our "psychospiritual immune system." These invisible processes help us resist the nonlocal mental intrusions of others, like our physical immune system guards us against infections. We might take a lesson in Era III from the premodern cultures in which nonlocal mind was taken for granted. In all these cultures, protective rituals abounded. Again, I have discussed how we can protect ourselves against negative nonlocal influences in my book *Be Careful What You Pray For.*[61]

CAN WE HAVE TOO MUCH ERA III?

In living an Era III life, we must not forget that we are also local creatures. There is always the need to stay in touch with our physicality. Even the great saints and mystics have complained about being too nonlocal, too often. A typical lament came from St. Teresa of Avila, who repeatedly experienced going outside herself and becoming one with God. She complained in 1577, "I've had raptures again. They're most embarrassing. Several times in public . . . during Matins, for instance. I'm so ashamed, I simply want to hide away somewhere!" In the same vein, Suzuki Roshi, who was influential in bringing Zen to America, said, "What do you want to get enlightened for? You may not like it."[62] Thus we see great spiritual teachers deemphasizing the epiphanies and intoxications of ecstatic unity in favor of something more ordinary, more *tolerable.* For example, when D. T. Suzuki, the legendary modern scholar of Zen Buddhism, was asked what it was like be enlightened, he chose not to speak of the rapturous component of this experience. Rather, he said, enlightenment is like ordinary awareness—except "about two inches off the ground."

Too much Era III can take other forms, such as the uncontrolled ability to see the future. This problem was brought home by a forty-year-old woman who came to see me for problems with rapid heartbeat, anxiety, insomnia, and headache. She revealed that all her life she had been able to see the future, a trait she shared with her mother and a sister. She worked as a radio dispatcher for the local police department. One day while at work she saw a tragic scene unfolding in her mind—a toddler falling into a swimming pool at an apartment complex somewhere in the Dallas metropolitan area. From past experience she knew that this event would happen in the near future and that she must prevent it. As a radio dispatcher, she had the power to send police immediately to the scene, but she could not identify the address. Soon after, she received a radio call from a police team who said they had pulled a small child from a swimming pool—too late. The woman felt she had caused the drowning. She was overcome with a sense of guilt and began to experience the symptoms of anxiety that brought her to my office. Initially she wanted only one thing—to get rid of her "gift." I referred her to a psychologist who knew about nonlocal mental states. With counseling and biofeedback training, her symptoms were soon under control. Her nonlocal way of knowing did not disappear, however, and she did not want it to. She began to see it in a new way, as a natural part of who she was. She accepted that her ability was imperfect and would always be so. She ceased making excessive demands on herself, and her guilt about the drowning incident resolved.

FEARS OF NONLOCAL MIND

As we consider applying Era III in our daily life, sooner or later we must ask the questions "Is this a good idea? Do I really want to experience nonlocal events?"

Psychologist Charles T. Tart, of the University of California, Davis, once asked a group of professional researchers in parapsychology to participate in a thought experiment. He invited them to imagine as intensely as possible, for ten minutes, that he had developed a new drug called telepathine. This marvelous substance would open up the mind so that anyone taking it would know all the thoughts and feelings of everybody within a hundred-yard radius. Moreover, the effects were permanent. At that point, intense intellectual discussion broke out in the group as the researchers discussed the impact of taking the new drug. "Now, who among you wants it?" Tart asked. There was not a single taker.

Tart was fascinated. Why would professional researchers in the field of parapsychology, who already believed in the capacity of the mind to behave nonlocally, be so leery? Tart continued inviting audiences to participate in the thought experiment. Almost everyone, laypeople and professionals alike, continued to reject taking telepathine. "I could be swamped by people's dreams. I could never get any sleep," one individual said. "If I could turn it on and off, choose what to know and *not* to know, it would be okay," said another. There were concerns about personal boundaries. "If you had somebody else's thoughts, how would you ever distinguish if they were theirs or yours?" one person queried. Others were concerned about the responsibility and power this would bring. One person said, "You'd know so much more, it might be hard just to be able to participate in society. You'd have insight for changes . . . , but suppose no one believed you or everyone thought you were crazy?" One participant told Tart she already has this kind of ability and has to be careful. She had the habit of telling people at work the answer to a question before they asked it. It got her into trouble, and she had to go to great lengths to censor her behavior.[63]

Some people equate Era III and nonlocal mind with losing their individuality, and they object to this on religious grounds. Following a lecture I once gave about nonlocal mind, a listener was upset. "In Christianity," she insisted, "the emphasis is on the individual. Jesus died for *my* sins. If what you say is correct, Christianity is wrong." I realized later that another view was possible from within her religion. In Christianity the part *does* yield to the whole and is gathered up into it, as in the Trinity. And sometimes the sense of individuality takes a back seat, as when John the Baptist said of Jesus, "He must increase, but I must decrease" (John 3:30 KJV). Catholic theologian Brother David Steindl-Rast has suggested that the fear of losing our individuality is unfounded. "Interdependence," he states, "which is joy of belonging and of being together . . . is what we really most want, but . . . we . . . are afraid to let go of . . . our independence and the isolation which necessarily goes with it. The moment we let it go, we die into the joy of interdependence."[64]

CHAPTER 5

REINVENTING MEDICINE

All these separations and gaps shall be taken up and hook'd and link'd together . . .
Nature and Man shall be disjoin'd and diffused no more . . .

WALT WHITMAN
Passage to India

Now that we have examined nonlocal mind, the question becomes how to put it to use in healing. Physicians, nurses, and health-care workers of every variety are already doing so; they literally are reinventing medicine. Let's further examine the background of these developments and the forms they are taking.

I. ERA III MEDICINE IN CONTEXT

ERAS I AND II: THE HEALING FOUNDATION OF ERA III

Our premodern ancestors used nonlocal mind in many ways—to know where to find game, when to plant, where to find lost objects, when to make war. But for them, as for us, the most dramatic use of nonlocal mind was probably health related. When a child, mate, or member of their family, tribe, or clan was sick, they applied their intentions, wishes, or prayers for healing. Today we are distracted from the healing powers of nonlocal mind because our attention is often fixed on the latest drug or surgical breakthrough.

The hold of physically based, Era I therapies over our ideas of healing is not difficult to understand. These methods brought hope and life where there were only death and despair. Consider, for example, the role these therapies played in dealing with infections. A little more than a century ago, ordinary folk in America had no idea that the diseases that often killed them and their children were due to microorganisms. The phrase "germ theory of disease" did not come into common use in the English-language medical literature until around 1870. In the late nineteenth century, no one was safe from the threat of mysterious fevers and plagues. "Even the most powerful and respected figures in Anglo-American society knew the humbling experience of deathbed scenes," writes Nancy Tomes, professor of history at State University of New York, Stony Brook, in *The Gospel of Germs: Men, Women, and the Microbe in American Life.* "For example, typhoid fever robbed a war-weary President Lincoln and his wife of their adored son Willie; left Queen Victoria grief-stricken over the death of her husband and the near-death of her eldest son; and smote the mother of a future president, Theodore Roosevelt, the same day that his wife died in childbirth."[1]

To understand the power that Era I continues to exert, let's take a snapshot look at what things were like a little more than a century ago in America.

The death at age forty-eight of Martha Bulloch Roosevelt, the mother of Theodore, Junior, on February 14, 1884, illustrates the fear felt by many Americans about the "filth diseases," as they were becoming called. Although no one knew the specific causal agents, these were illnesses known to be spread by fecal contamination. As such, they were considered preventable by proper domestic sanitation. By all accounts, Martha Roosevelt, known familiarly as "Mittie," was so concerned about cleanliness that her friends thought her almost obsessive.

The mansion the Roosevelts occupied on West Fifty-Seventh Street, built by Theodore, Senior, in the 1870s, had been designed by the leading architects of the day and incorporated the latest advances in fixtures and appointments. Mittie had moved from rural Georgia to New York City after her marriage in 1853 and was appalled at the extraordinary filth of the rapidly growing city. Special measures were required to combat it. These included a daily bath with two changes of water, spreading a sheet on the floor at night when she said her prayers, and wearing white clothing even in winter to make sure that no speck of dirt escaped detection. Assisted by her small army of servants, she kept her house as clean as herself. She compiled a lengthy set of housekeeping instructions, including a rigorous program of polishing, scrubbing, sweeping, and dusting. When the ash man came each morning, for example, the cook was required to greet him with a bucket of scalding water to decontaminate his ash can before he could enter the house.[2]

"Yet for all these heroic efforts to keep herself and her home spotlessly clean," Tomes writes, "Martha Roosevelt succumbed to what public health experts knew to be a 'filth disease.' Such a fastidious woman would surely have died of shame, had the typhoid fever not killed her, to discover that she had contracted a disease spread by fecal contamination. Martha Roosevelt's death from typhoid epitomized the uncertainties that beset even the most conscientiously clean households of the Gilded Age: no one, not even the most careful, seemed to be safe from the invisible agents of disease."[3]

Rates of illness and death rose sharply in the large cities of America and Europe in the nineteenth century, making it dangerous to live in them. Urban dwellers endured not only epidemics of cholera and smallpox, but also endemic diseases such as typhoid and pneumonia that killed at a steady rate, year after year. In the 1880s in New York City, one-fifth of all infants died before reaching their first birthday, often of infant diarrhea, and even those who reached adulthood had a one-in-four chance of succumbing between the ages of twenty and thirty.[4]

Infections were a way of life, and all the social classes in nineteenth-century America were good at recognizing them, as we saw in chapter 1. The result was a chronic sense of profound dread. One simply could not escape death from infections; not even the "best" households were safe, as the Roosevelt experience showed.

Then, as widespread public health measures were introduced in the late nineteenth century, disease rates plummeted. It must have seemed miraculous. By the time vaccines appeared in the early twentieth century, and sulfonamides in the 1930s and penicillin in the 1940s, the healing benefits of Era I medicine had already begun to seem limitless. Thus by 1962, Sir F. Macfarlane Burnet, the Australian immunologist and Nobel laureate whose research paved the way for organ transplantation, was moved to say that the twentieth century would witness "the virtual elimination of infectious disease as a significant factor in social life." To write about infectious disease, he observed, "is almost to write of something that has passed into history." Seven years later in 1969, William Stewart, the United States Surgeon General, testified before Congress that it was time to "close the book on infectious diseases."[5]

Today we know that these predictions were premature; along with the appearance of new diseases such as AIDS, antibiotic-resistant strains of bacteria are appearing worldwide, and experts are no longer sanguine about our ability to hold them in check. In spite of these challenges, the advances against infections, along with breakthroughs in public health, surgery, and anesthesia, heralded a golden age of health care compared to what had come before.

When physically based approaches seemed so powerful, why pay attention to the powers of the mind, as a few researchers began to do in the 1950s? Many Era I proponents believed that it was simply meddlesome to introduce emotions, feelings, and thoughts into the medical equation. Even if these factors had effects on health, they were surely trivial when compared to physical factors such as microbes, genes, and sanitation. Yet the power of the mind kept hammering for admission into the house of healing, and finally it gained entry. One door was the placebo response—the effects of belief and expectation on the actions of drugs and surgical procedures. These effects were so significant that special "double-blind" research methods were devised to take them into account. They required that a control group be compared to a treatment group and that no one, neither subjects nor experimenters, know who is receiving the drug being tested and who is receiving the inactive placebo. Double-blind protocols soon became the golden standard in drug research.

But emotions were soon found to be more than a factor in medical experiments; they were discovered to make a life-or-death difference in specific diseases. The Type A personality—an individual who is burdened with a chronic sense of urgency, who always feels he or she is running out of time—was described in the 1970s by cardiologists Friedman and Rosenman. These individuals were shown to die more frequently and earlier in life from heart attacks. Certain emotional states were demonstrated to make a difference *after* a heart attack, during hospitalization in the coronary care unit. Social contact was proved to be an important factor in resisting all major diseases. Emotionally stressful events such as the death of a spouse, changing jobs, and even taking a vacation were found to correlate with health problems. Job dissatisfaction was found to be a major predictor of heart attack. A loving relationship with one's spouse proved to be as effective as drugs in reducing angina, the pain of heart disease. The absence of loving, close relationships with parents proved a powerful predictor of cancer in medical students. Bereaved patients were shown to have immune systems that practically shut down during bereavement. Relaxed states, such as during meditation and prayer, were found to lower blood pressure and heart rate and to create a generally healthy bodily response.[6]

As a result of these and many similar discoveries, entirely new areas of medicine began to form, such as the field of biobehavioral cardiology, the use of specific behaviors to prevent and treat heart disease, and psychoneuroimmunology, which recognizes links between the brain and the neurological and the immune systems. Mind-body research exploded and brought with it new therapies based on the new findings. Imagery and visualization techniques were developed by O. Carl

Simonton, Stephanie Simonton, Jeanne Achterberg, G. Frank Lawlis, and others. Meditation began to be used to lower blood pressure and cholesterol levels in the blood.[7] Biofeedback, in which sophisticated electronic instruments monitor the physical changes associated with one's thoughts and feelings, proved helpful in controlling migraine headaches, irritable bowel syndrome, anxiety, insomnia, and other stress-related problems. In less than two generations, the mind-body medicine of Era II was firmly positioned alongside the physical approaches of Era I as a legitimate part of scientific medicine.

In a sense, Era II "reinvented" Era I; that is, Era II demonstrated the incompleteness of Era I and expanded its scope. But it did not replace it. The terms *complementary* or *integrative* medicine have arisen to describe how these methods are being used in conjunction with one another—a both/and, not an either/or, approach.

In just the same way, Era III is reinventing and expanding both Eras I and II—not replacing them—in ways we shall examine shortly.

Now as we examine how Era III is changing modern medicine, let us bear in mind the features that set it apart from Eras I and II. Era III views consciousness as unconfined to the physical body, as infinite in space and time. In Era III, consciousness works *through* the body but is not limited to it. In the Era III view, consciousness cannot be explained in terms of the physical processes taking place in the body. This means that consciousness is *fundamental*—not derived from anything more basic, not explainable by anything more elemental. Era III medicine, therefore, is *a form of healing based on the fundamental, infinite nature of consciousness.*

II. ERA III MEDICINE AND THE PHYSICIAN

A. ERA III MEDICINE AND DIAGNOSIS

Two of the main pillars of modern medicine are diagnosis and treatment. In Era III, both are subtly but profoundly altered.

THE COMPUTER AND THE INTERNISTS

In a research project at a major medical school, a computer was programmed to interview patients and diagnose their illness. The accuracy of the computer was then compared to the performance of a group of skilled internists. The computer and the physicians asked patients identical questions, and the patients gave identical responses to both of them. The computer, however, was not nearly as successful as the doctors in making a correct diagnosis. The researchers were puzzled why this should be so. So they asked the internists, "What is the first thing you notice when you begin interviewing a patient?" They replied, "Whether or not he or she looks sick." The researchers could not objectively determine what "looks sick" meant and therefore could find no way to program "looks sick" into the computer. As a result, the research project was abandoned.

Any skilled diagnostician, however, knows what the internists were doing. They were going beyond factual information and yes-and-no answers. They were bringing into play what is sometimes called a gift—the ability to "just know" or to *intuit*—which, I contend, involves nonlocal ways of knowing.

Intuitive, Era III knowing is not a sequential, step-by-step process of deduction in which we reason from *a* to *b* to *c*. It is an all-at-once phenomenon in which information presents itself whole, fully formed. It is more like revelation than reason. When we know something intuitively, we know it suddenly and completely. Nonlocal knowing is often a surprise; we feel clobbered by a conclusion we didn't see coming.

Diagnosis is derived from Greek words meaning "through" or "between" and "know." These root words suggest intuitive knowing—opening our perceptions to answers that already exist, which are woven *between* the data and *through* the facts. Making a diagnosis is like confronting a tapestry. Our task is to absorb the pattern, to grasp the design, not to bludgeon our way through it with linear reasoning.

Among my internist colleagues there is a saying: "If you don't know the diagnosis one minute after a patient starts to tell her story, you're in trouble." This reflects the possibility of immediate knowing without lengthy interviews and expensive diagnostic workups. It represents intuition in action, the Era III way of doing things. I've seen diagnosticians so skillful that I am awed in their presence. They seem to be more magician than doctor. Even when I've asked them to explain their line of reasoning in a particular case, they are often unable to do so. I'm convinced they often employ nonlocal ways of knowing without being aware of it. Yet we resist this possibility. Nonlocal mind is so out of favor in medical circles we camouflage it with various aliases—a lucky guess, a hunch, "the obvious." But these terms only increase, not resolve, the mystery of how physicians often arrive at their conclusions.

I'm not saying that doctors don't need logic and reasoning, only that there are times when their conclusions cannot be predicted by the facts they have to work with. A physician who uses nonlocal mind *and* logic will be more effective than one who uses either alone.

Nonlocal ways of knowing have been cropping up in the medical profession since the Age of Enlightenment. Like many medical developments in their early stages, they often have a melodramatic, soap-operatic quality. Let's take a look at part of this rich, colorful history in which nonlocal mind is struggling to be born.

SNAP DIAGNOSIS

During the eighteenth century the practice of "snap diagnosis" spread throughout the best medical schools in Europe. The trick was to make a diagnosis with as few clues as possible and to rattle it off immediately. Eminent physicians vied with one another in this endeavor, which often involved considerable showmanship. One of the best snap diagnosticians was Napoleon's favorite doctor, Jean-Nicolas Corvisart, who made contributions to the art of physical diagnosis and the understanding of heart disease. Corvisart once added to his reputation by merely looking at the subject of an oil painting and correctly diagnosing him as suffering from cardiac disease. Hans von Hebra and Joseph Bell, other well-known physicians of the day, could discern not only the diseases of their patients but their occupations as well. The famous German physician Friedrich Theodor von Frerichs was so infatuated with his powers of snap diagnosis that he never admitted a diagnosis to be wrong. Frerichs could be theatrical, and he was worshiped by his students, who, it is said, "hung on his lips, and . . . revered his wonderful precision."[8]

This sort of thing was not new, of course. Shamans, visionaries, and folk healers had done it for centuries. But that snap diagnosis should crop up during the eighteenth century, when medicine was becoming increasingly scientific, seems decidedly odd.

MESMERISM AND PSYCHIC DIAGNOSIS

In Europe, psychic or clairvoyant diagnosis had captured attention since 1776, the year Franz Mesmer, a young medical student at the University of Vienna, wrote his dissertation on what would be known as "animal magnetism" or mesmerism. Although mesmerism and psychic diagnosis were embraced by the public, they had tough sledding among the scientists of the day. Louis XVI convened the Royal Commission of Inquiry of 1784, headed by the naturalist Benjamin Franklin, the United States ambassador to France, to evaluate these developments. No sooner had the commission denounced them than fresh evidence for psychic diagnostic abilities began to pour in.

Some of the reports came from Amand-Marie-Jacques de Chastenet, Marquis de Puységur (1751–1825), who stands second only to Mesmer in the annals of mesmerism. A wealthy French aristocrat, he was well educated in literature and the sciences. He became interested in animal magnetism when his asthmatic brother, whose severe illness defied the efforts of the best doctors in Paris, was cured over a three-month period by Mesmer. One of his most famous patients was a twenty-three-year-old peasant, Victor Race, who suffered from a severe pulmonary condition characterized by fever, pain, and coughing up blood. He improved dramatically with Puységur's mesmeric treatments, which included Race's entry into deep trance states. During his patient's convalescence, Puységur discovered that Victor possessed some remarkable gifts. When placed in trance, Victor became clairvoyant. As British historian Alan Gauld states, not only was Victor able to diagnose and prescribe for his own illness and predict his course, "on being brought into contact with other patients, he seemed able to do the same for them."[9]

In 1813, Joseph P. F. Deleuze (1753–1835), a French naturalist and librarian, published a few cases in which people could diagnose disorders in themselves and others when mesmerized. In 1818, Professor D. Velianski, professor of physiology and pathology at the Imperial Academy of St. Petersburg, drafted a detailed description of the mesmeric trance. He described how, as the trance deepened, individuals could often "see" into their own and other bodies, make diagnoses, and recommend a course of treatment. Dr. Alexandre J. F. Bertrand (1795–1831), a Paris physician, mathematician, and skeptic on such matters, discovered that a few of his mesmerized patients had the faculty of diagnosis, as if by "second sight," and could tell him what was wrong with his other patients. One individual was able to do this by experiencing the symptoms of the patient in her own body, and she occasionally picked up disorders Bertrand had missed. On one occasion Bertrand asked her what was the matter with one particular man who was to arrive shortly at his office. The woman correctly informed him that the man had been shot, although no wound was visible and Bertrand was unaware of the incident. Justinus Kerner (1786–1862), a district medical officer, poet, and writer in Weinsberg, Germany, described the feats of Frau Friedericke Hauffe (1801–1829), the famous "Seeress of Prevorst." From an early age she was subject to prophetic dreams and visions. At age twenty-one she became quite ill and seemed to be dying. Kerner treated her with mesmerism and found that when she was in an altered state of awareness she became clairvoyant and could diagnose and prescribe for herself and for others. In several cases her predictions and visions were verified.[10]

In 1826, due to the increasing evidence, the French Academies of the Sciences and of Medicine agreed to appoint another commission of inquiry. The skeptics, who controlled the proceedings of the commission, were given a shock by Céline, a patient of Dr. P. Foissac (1801–1884?), a young Paris physician. Although totally ignorant of medical matters, she told the members of the commission what was wrong with three patients. Moreover, she recommended particular therapies, which she could hardly have known about. One patient was already being treated with mercury for what was called a venereal infection. Céline implied that the medication and not the disease was causing the symptoms. Her advice was ignored, and the patient died shortly thereafter. An autopsy revealed she was correct and the doctors were in error.[11]

Some of the psychic diagnosticians even provided evidence that the materialists suppressed, since the skeptics believed, then as now, that such things were impossible. In 1839 the French physician G. P. Billot described how he had discovered a subject who when mesmerized could diagnose illness in patients and prescribe for them as well. Once, when she prescribed a flowering herb not in season, the plant suddenly materialized. This had happened twenty years earlier; Billot had deliberately not reported such an outrageous event. He overcame his hesitation when his correspondence with Deleuze revealed that Deleuze, too, had witnessed similar examples but also thought it unwise to publish them. "I have suppressed many things in my works because I considered it was not yet time to disclose them," he said.[12]

What are we to make of these happenings? Compared to modern experiments such as those reviewed in chapter 2, these reports seem quaint. Standards of observation and reporting were crude at the time, and no one is justified in basing a case for nonlocal mind on events such as these. But perhaps we can be too critical. As psychologist Alan Gauld states in his exhaustive history of hypnosis,

> One must not be misled into supposing that these shortcomings are in all cases sufficient to justify us in totally dismissing the claims of the enthusiasts. Often the enthusiasts are on to something, even if they haven't got it quite right. The animal magnetists were certainly on to *something*, and it is debatable whether the progress we have

> made in the intervening centuries is such as to justify us in being patronizing towards them.[13]

PSYCHIC DIAGNOSIS AND SPIRITUALISM

Americans, too, experienced similar developments. One of the most famous psychic diagnosticians in recorded history was Mrs. W. R. Hayden, married to a former editor of a New England newspaper. She was a product of "spiritualism," a movement that swept America by storm in the mid–1800s and was the rage of cultivated folk everywhere. Spiritualists, or mediums, claimed they could communicate with the spirits of the dead, make tables levitate, create unaccountable sounds, and cause musical instruments to play on their own. It is difficult today to realize how seriously spiritualism was regarded. We get a hint from the list of the individuals who, in 1850, visited the Fox sisters of Hydesville, New York, near Buffalo, who started the craze in 1848. The list included Charles Dana of the *American Encyclopedia;* Charles Bancroft, the historian and diplomat; John Bigelow, editor of the *New York Evening Post,* later U.S. minister to France; William Cullen Bryant; and James Fenimore Cooper. Arthur Conan Doyle, the champion of rational deduction who gave us Sherlock Holmes, was sufficiently impressed with the movement to write a *History of Spiritualism.*

It didn't take physicians long to investigate. The newly formed Buffalo School of Medicine sent three professors to investigate the Fox sisters, who were famous for séances during which strange sounds were heard. When the doctors discovered that the knocks did not occur when the women's knees were held, they figured they had cracked the case: the sounds were obviously caused by the women's ability to pop their knee joints. This explanation was dealt a severe blow, however, when it was realized that the sounds occurring in the sisters' séances were wildly varied. Not even the Buffalo doctors could explain how they could produce sounds such as sawing wood and clanging hammers from their joints and how tables could levitate and heavy objects float through the air without any demonstrable interference.

Unlike the Fox sisters, Mrs. Hayden was widely regarded as genuine. In 1852 she traveled to England and introduced the British to spiritualism. As befitting the British ambience, her séances were staid; typically the only phenomena to which the Brits were treated were sounds or "rappings." The most detailed accounts of Mrs. Hayden's séances were provided by Sophia and Augustus de Morgan, the first holder of the chair of mathematics at University College, London, and the secretary of the Astronomical Society. De Morgan was skeptical of mediumship, but at Sophia's insistence he decided to get involved. He arranged for séances to be conducted in his own home so he could control the setting. Mrs. Hayden would come alone so there could be no collusion between her and someone else. Many tests of her ability "to know" were conducted. De Morgan would ask questions of her, not by writing them on paper or asking them aloud, but by *thinking* them. He was stunned by the accuracy of her responses. Although de Morgan was never convinced that Mrs. Hayden had made contact with "spirits," he stated in the end that he was "perfectly satisfied that something, or somebody was reading my thoughts."

Mrs. Hayden was never detected in any deception in spite of the monumental efforts of scoffers on two continents to expose her as a fraud. She returned to the United States and went to medical school. She became a physician, which was no small accomplishment for a woman in those days. She is one of the first psychic diagnosticians America produced. She went to work for the Globe Insurance Company, apparently on the strength of her psychic diagnostic abilities. Joseph R. Buchanan, a neurologist interested in the psychic aspects of medicine, described Dr. Hayden as "one of the most skillful and successful physicians I have ever known."[14]

RECENT DEVELOPMENTS

The old powers still lurk, and nonlocal diagnosis is currently enjoying a renaissance.

In several popular books, Carolyn Myss, a former publisher, describes how she has had psychic abilities since childhood. Several years ago she began to work with Norman Shealy, a Harvard-trained neurosurgeon, in psychically making diagnoses of his patients. Shealy put her to a test. With a patient in his office in Springfield, Missouri, he would phone Myss in Walpole, New Hampshire, and provide her with the patient's first name and birth date. Myss would then provide Shealy with the patient's diagnosis. In the first 100 cases, Shealy, who is no stranger to research methods, claims she was 93 percent correct.[15] Such a robust claim, of course, demands replication under controlled conditions before being accepted.

Judith Orloff, M.D., a California psychiatrist in private practice, has recently described her use of psychic diagnosis in her book, *Second Sight.* Interestingly, Orloff's abilities flowered following two near-death experiences, but she contends that intuitive abilities are innate in everyone, physicians and laypeople alike. "We are all visionaries," she states. "Even if you don't think of yourself as psychic, our prescience lies latent, a shared legacy we each have the right to claim. That any of us have ever been forced to suppress our psychic experiences is a travesty. . . ." For Orloff, the value of acquiring knowledge nonlocally goes beyond making accurate diagnoses. "Though it can be simply a means of information gathering," she says, "I've found its highest value is in penetrating the layers of reality that reveal the interconnectedness of all things . . . with yourself, with others and with the world around you . . . [and] with spirit."[16]

The most recent physician to write about her experiences with psychic diagnosis is neuroscientist and neuropsychiatrist Mona Lisa Schulz, M.D., Ph.D., in her book, *Awakening Intuition.* Like Orloff, Schulz believes that intuition functioning nonlocally, beyond the body, is natural in everyone. "If you have a brain and a body," she states, "if you have memories, if you sleep at night (or any other time), then, by definition, you have to be—*you are*—intuitive."[17]

Schulz began to put her intuition to work as a medical student. During her first day on duty at a very busy, poorly staffed, inner-city hospital, she was told to go down to the emergency room to see her first patient, a fifty-six-year-old woman named Betty. She was to find out her complaint, take her history, and do a physical exam. As soon as Schulz heard the patient's name, she found herself envisioning what was going on in Betty's body. At the same time, she saw problems in Betty's emotional life that might have contributed to her physical problems. The issue seemed to be some family responsibility she was ambivalent about discharging. Schulz visualized Betty as around five feet four inches, obese, and suffering from severe pain in the right upper part of her abdomen, perhaps from her gallbladder. When Schulz arrived in the emergency room, it was as if her intuition had sprung to life. Betty was a middle-aged, overweight woman clutching the right side of her abdomen. She had been on her way to a family reunion and was quite upset that she would disappoint her family if she was not able to attend. Schulz ordered the appropriate tests, which proved her correct: Betty had gallstones and an inflamed gallbladder.

"Using my intuition helped me organize myself better in the hospital," Schulz observes. "I ended up being faster, more efficient, and able to leave the hospital sooner than others. Moreover, I was using my intellect *and* my intuition in a productive and fulfilling fashion." She concludes, "My intuition, working with my intellect, made me a better physician."[18] Schulz refined her intuitive skills through continual practice through the years. Today, she can't imagine functioning without invoking them.

"The only mention of premonitions or other psychic abilities I ever found during my medical education was in textbooks labeling such claims as a sign of profound psychological dysfunction," says Orloff. In keeping with this view, Orloff tried to deny her intuitive skills because they felt "alien" and "threatening." Yet, try as she might, she could not extinguish her ability to know nonlocally what was wrong with her patients and what would happen to them. Once, she "knew" that a patient of hers, who appeared to be doing well clinically, was going to attempt suicide. Orloff disregarded this vision; her patient indeed tried to kill herself and nearly succeeded. She was in a coma in the intensive care unit for several weeks. As a result of denying her premonition, Orloff felt "both validated and terrified. . . . I suddenly felt as if two distinct parts of me had collided." It was a telling moment in her life. She realized that *not* using her intuition placed her in an ethical and moral dilemma, because these skills might make a life-or-death difference for her patients. She gradually made peace with her ability, becoming an Era III physician in the process.[19]

NONLOCAL DIAGNOSIS BY PATIENTS

Patients, too, appear to "just know" what is going on in their bodies. In a 1991 study conducted at the Department of Epidemiology and Public Health at Yale School of Medicine, researchers Ellen Idler and Stanislav Kasl studied the impact of peoples' opinions on their health. The study involved more than 2,800 men and women aged sixty-five and older. Their findings are consistent with the results of five other large studies involving more than 23,000 people, ages nineteen to ninety-four, of both sexes. All these studies came to the same conclusion—that the best predictor of survival over the upcoming decade is the answer people give to a simple question: "What do you *think* about your health?" The predictive power of the answer to this question exceeded objective factors such as physical symptoms, extensive exams, and laboratory tests, or behaviors such as cigarette smoking. For example, people who smoked were twice as likely to die over the next twelve years as people who did not, whereas those who say their health is "poor" are *seven* times more likely to die than those who say their health is "excellent," all other factors being equal.[20] This is one of the most potent predictors of illness ever discovered. How do people know what lies ahead? A mind-body, Era II process may be occurring, in which they tune in to messages from their body, detecting signals too subtle to show up on physical examinations and tests. In addition, they may be participating in Era III, nonlocally scanning the future and detecting what lies ahead.

Using these findings, Era III physicians would routinely ask people during periodic health examinations what they think about their health. If they give a negative answer, the physician would take this response seriously, even though the physical exam and lab tests might be normal. The doctor might encourage the patient to attend to his or her health more closely than usual and perhaps report for more frequent examinations.

It is also likely that people can sense health problems nonlocally in others. When a mother "just knows" her baby is ill, in spite of nothing objective to go on, and her hunch is confirmed by her pediatrician, she may be invoking nonlocal ways of knowing. This possibility has been affirmed in sudden infant death syndrome (SIDS), in which parents often have premonitions of their child's death (p. 119). When a busy executive has a hunch that she should set her busy schedule aside and go in for her annual exam ahead of time, and discovers a health problem in its early stages, she may also be employing nonlocal mind.

THE NEED TO USE ALL THREE ERAS IN BALANCE

I would not submit to brain surgery just because a psychic diagnostician told me I had a brain tumor. When serious illnesses are involved, multiple ways of making a diagnosis should be used, because no method of diagnosis is always correct.

Critics of Era III often paint lurid scenarios in which doctors abandon science, jettison reason, and rely solely on intuition, bumping off thousands of patients in the process. There is no evidence whatsoever that physicians will desert science as they become familiar with the evidence favoring nonlocal mind. One reason is that the case for nonlocal mind *is based in science.*

The contention that Era III methods will result in wholesale slaughter is overwrought. Surveys show that the use of prayer and spiritual healing, for example, are not associated with neglect of conventional therapy.[21] Although there are indeed religious groups who believe in the exclusive use of nonlocal, spiritual methods in healing, the vast majority of Americans favor using all three approaches together, *combining* the physical, mind-body, and nonlocal approaches of Eras I,

II, and III. When it comes to choosing therapies, *instead of* is virtually absent from the American vocabulary. This reflects a thoroughgoing pragmatism lying at the heart of the American character—the belief that one should cover *all* one's bases, throw the book at the problem, use what works.

Balancing Era I and Era III can also help people make sense of their experiences when they are sick. An example involves cancer patient Claire Sylvia, who underwent a successful heart and lung transplant at Yale–New Haven Hospital in 1988. In her memoir, she documents how, having never before liked foods such as beer and chicken nuggets, she suddenly started craving them following her transplant. She tracked down the family of her donor—a teenaged boy, Tim, who died in a motorcycle accident—and was able to verify that her new food tastes and even her dreams were similar to those experienced by Tim while he was alive.[22] After Sylvia's story, *A Change of Heart: A Memoir,* was published and her experiences were featured on *60 Minutes,* other transplant recipients have come forward with similar stories.

Thus far, almost all the explanations of Sylvia's experience have centered on two possibilities. One is coincidence ("just one of those things") or wishful thinking (seeing what you want to see), which is the predictable skeptical response. The other is the "cellular memory" hypothesis, the possibility that our thoughts and desires are somehow lodged in our very tissues. This implies that if our organs are transplanted into someone else, something of our emotional life is transferred along with them, which might be experienced by the recipient. Cellular memory is a "cassette theory"—the donated organ is the cassette tape containing the information, and the recipient's body is the cassette player decoding and playing it back.

The "transplant phenomenon" is being embraced in our culture as something new, but the theme is ancient. It is a variation on reincarnation—the idea that someone who has died comes to life again. The new twist is organ transplantation, which did not exist in the annals of reincarnation until now.

If an organ is removed and inserted into someone else, Eras I and II thinking suggests that something of the mind may be transferred along with it. Yet although our cells and tissues can be affected *by* thought, there is no evidence that specific thoughts can hitch a ride on them when they are transplanted into another individual. Were this not true, the millions of people who have had blood transfusions and skin grafts—both forms of organ transplantation—would almost certainly have noticed this phenomenon long ago. The basic mistake, in my opinion, is in equating physical tissue with thought itself. This is a common trap we fall into, as when we equate the brain with consciousness. Yet this is like saying the television set actually produces the image on the screen. Far more likely, in my judgment, is that our cells and organs function to *receive* and *transmit* thought. Again, the TV analogy fits: the television set receives a signal from an outside source; it does not manufacture it.[23]

It is easy to see how we might be misled on this point. Imagine a battery-powered television set plopped down in a remote jungle, where the inhabitants have had no prior contact with modern culture. When the TV is turned on and images of people appear on the screen, the natives would think the images originated in the box itself. How could they think otherwise, with no concepts of invisible electromagnetic signals being bounced off communication satellites orbiting in space?

In the same way, when we witness an organ transplanted into a body, and all of a sudden new images, dreams, and traits play themselves out in the recipient, it is natural to think that they, too, originated in the transplanted organ. Yet there is no evidence that specific thoughts reside in a transplanted organ, just as there is no evidence that the image on the TV screen lives in the machine itself.

Era III thinking suggests that the consciousness of the organ donor is nonlocally united with the consciousness of the recipient. It is this connection, I propose, that enabled Claire Sylvia to gain information about her donor. Receiving the donor's heart did not actually transfer the experiences mechanically from him to her but instead somehow intensified a mental connection that was already nonlocally present.

B. ERA III NEW PROTOCOLS AND TREATMENT

"The toughest test for nonlocal healing is the emergency room," a colleague of mine recently remarked. "If it works in the ER, it'll work anywhere." I agree with him. So, in the following account, I've tried to imagine what Era III medicine would look like in the emergency room of a major hospital.

The following story is a composite. In developing this account, I have relied on proven science. Every feature I describe is based on solid evidence, from the effects of distant intention to mind-to-mind communication between health-care professionals. Although I know of no hospital that currently features *all* the elements I describe, many feature individual therapies I discuss, and some are coming close to combining all of them.

A PREVIEW OF ERA III IN THE HOSPITAL: THE STORY OF ROSEMARY GREY.

While returning home one wintry night, Rosemary Grey's car skidded on a patch of ice and struck a tree. When the ambulance arrived she was lying unconscious and bleeding heavily but alive. The emergency medical technicians started intravenous fluids, stabilized her head and neck to prevent spinal cord injury, and rushed her to Mercy Medical Center.

As Rosemary was being transferred from the ambulance to a stretcher, the orderly noticed the distinctive bracelet she was wearing. It was similar to those worn by patients to indicate drug allergies and underlying medical conditions, except that it was engraved simply with "III." This was the orderly's cue to wheel Grey into an Era III trauma room.

The treatment of physical trauma was one of the great achievements of twentieth-century medical science. Mercy Medical Center had been in the forefront of these developments. Now it was leading the field in Era III trauma care, which incorporated the latest advances in Era III healing. Research had established that the effects of drugs and surgical procedures could be enhanced by the thoughts and intentions of caring, compassionate individuals. But although these discoveries were widely recognized and had already been integrated into hospital wards and outpatient clinics, Mercy was one of the few hospitals that had carried nonlocal methods into the emergency rooms. Most institutions considered the ER off limits to Era III approaches. Where broken bones and hemorrhaging wounds are concerned, critics said, only physical, or Era I, approaches mattered. In the emergency room one had to act quickly and physically. This was not the place to try to *think* a patient healthy. Since the emergency room attracts physicians and nurses whose beliefs reflect this view, many who worked there were openly cynical about nonlocal healing, which they considered a nuisance or a passing fad. Soon, they hoped, the obsession with the mind would abate, and *real* doctors could get back to practicing *real* medicine.

Era III medicine, however, turned out to be no fad. Training programs in Era III medicine were being offered around the country, and board certification in this field was on the horizon—developments that had long fascinated Catherine Pierce, M.D.

Cathy Pierce was on duty when Rosemary Grey rolled into Mercy's emergency room. Within moments, Pierce had performed an Era I physical exam, employing her senses of touch, sight, and hearing. At the same time, she gave her intuition free rein. She entered Grey's body with her mind and scanned where her physical vision could not go. She was gifted in intuitive as well as in physical diagnosis, and she combined these approaches every time she encountered a patient. Pierce concluded that Rosemary Grey had been fortunate. Although still unconscious and covered with bruises and a nasty laceration across her brow, she seemed to have escaped internal injuries.

After a decade as a trauma physician, Pierce was so proficient she could perform most of her duties automatically and with great confidence. This allowed her the freedom to direct a considerable degree of her mental efforts to the practice of Era III medicine, such as trying to guide her patient's outcome with her intentions, wishes, and prayers.

In so doing, Pierce made her own rules. She rejected the popular approach of trying to *make* things happen with her intentions. In such matters Cathy Pierce was immensely fond of subtlety. She presumed that the world knew perfectly well how to get on without any advice from her. She believed that her task was not to dictate terms to the universe or overpower it with her thoughts; rather, she wished merely to nudge things in a certain direction through a bit of delicate mental tweaking.

She had not always felt this way. At first she had tried to *force* things to happen in her patients, certain that she knew best. While this sometimes worked, she found it unsatisfying because it felt arrogant and presumptuous. After years of experimentation, she concluded that her goal should simply be to set the stage for a wisdom greater than her own, then step aside. To some, this approach—healing by invitation, not command—smacked of passivity. She knew otherwise. Whether she called this benevolent power God, Goddess, or something else did not concern her in the least. Pierce's primary concerns were not theological.

She considered herself deeply spiritual, although not religious. But for her, spirituality was profoundly practical; it involved a sense of connection with something outside herself that transcended the human level of existence. This unnamed reality was more powerful than the individual self. She had seen too much evidence of her relative powerlessness as a doctor. She knew all too well the limits of her practice and that a greater force was at work when healing was involved. She respected whatever it was that made people get better after the incisions were made and the antibiotics given. So this "thing" was a continual presence, an inner compass that guided every aspect of her life.

Dr. Pierce could not separate science and spirituality. She felt a profound reverence toward the patterns and processes that science revealed. The filigreed order she saw in chemistry, physics, and mathematics was so intricate and astonishing, it made her shudder. And Dr. Pierce considered it her duty to honor the revelations of science wherever they might point and however profoundly they might challenge her prior beliefs.

When she first encountered data from actual experiments indicating that Era III healing worked, however, she was disturbed. Over the years, she had become comfortable with the idea that one's *own* mind could influence one's health, but she was not yet prepared to grant that one's mind could influence the health of another individual who was unaware of the effort. She did not retreat from these discoveries, though, when she read about them and witnessed them. As a physician, she could not disregard the practical implications of the studies favoring nonlocal healing. If the findings were valid, as she believed they were, and if she were to withhold these methods from her patients, she would be ignoring her commitment to put her patients' welfare first—ahead of her own biases or doubts.

Dr. Pierce was familiar with the nuances of the debates about Era III medicine. She found the arguments tiresome. She was convinced that if this healing were a new drug, it would be adopted by her profession overnight. But Era III healing was in many ways better than a drug. It was free, it could be used alongside conventional therapies, and it could be administered by anyone. You didn't need a white coat and a stethoscope to be a specialist in nonlocal healing; you had only to care, to love, to be compassionate.

You didn't even need to be religious. Pierce was delighted with this particular feature of the experiments in Era III medicine—that their outcomes were not correlated with any particular religion. Intuitively she felt that this finding was on target.

The bottom line of the healing experiments was that the compassionate intentions of one individual could help heal another, outside his or her awareness, no matter how far away the two might be. Pierce smiled when she considered how ancient this realization was. For folk healers throughout history this was Basic Healing 101. Where nonlocal healing was concerned, modern science didn't look so modern. Better late than never, she told herself.

She remembered the derision she encountered when she entered the two-year training program in Era III medicine fresh out of her residency in ER medicine. The concept of nonlocal mind was new at the time, and many of her colleagues were confused about what it implied. They often teased her, called her Psychodoc, but she went her own way. And when controlled studies showed that emergency patients treated with conventional *and* nonlocal methods did better than those treated with conventional methods alone, the jests faded.

During her college years Pierce had participated in a series of experiments that helped lay the groundwork in nonlocal healing. She had minored in microbiology, and one summer she worked as a research assistant for a professor who was using microbes as "targets" in some controversial experiments. The goal was to determine whether individuals could influence the growth rates of bacteria by "intending" them to grow faster than controls. The outcome of the studies was profoundly positive, and the results were soon replicated by other researchers.

The following summer her research moved to higher forms of life. She studied the impact of nonlocal thought on the recovery of laboratory mice from infections. Shortly thereafter, other researchers extended her positive findings to humans by showing that antibiotics *plus* nonlocal healing intentions were more effective than antibiotics alone.

Pierce now knew the research in the field of nonlocal medicine inside and out. She was amazed that many physicians ignored the evidence. Their reasons, she suspected, were rooted in the tendency of humans to hang onto what is familiar and comfortable. Many physicians, for example, continued to do coronary bypass surgery the old way, opening up the chest cavity, instead of using the new percutaneous fiberoptic methods that made open-chest surgery unnecessary.

It was not only professionals who resisted Era III medicine. A few patients did also, and they were willing to go to court about it. Legislation was already on the books requiring a patient's consent before a physician could use nonlocal healing methods on his or her behalf. Sometimes consent was impossible, as when a patient was unconscious, as in Rosemary Grey's case. This was why Era III bracelets had caught on. They were a way of authorizing a doctor to administer Era III care in advance, in case a patient was unable to give instructions.

Long before completing her training, Catherine Pierce had begun to consider where she would practice. She selected Mercy Medical Center because of its openness to innovation. Mercy had been one of the country's first institutions to integrate Era III techniques into its hospital wards. Was the hospital also willing to carry Era III medicine into the emergency room? This was Pierce's key question during her job interviews. She knew this was asking a lot. Nonetheless, the hospital administrator and the physician chief of staff agreed to back a study to determine whether nonlocal approaches made a difference in treating emergency cases. For three years, half the patients coming through Mercy's ER would be treated conventionally, and half would be treated conventionally *and* with nonlocal techniques. Pierce was offered the opportunity to direct the study. The experiment had validated the nonlocal approach, an outcome she never doubted.

The actual practice of Era III medicine had not come easily for Pierce, however. She found it difficult to perform all the technical functions in a crisis and keep her mind in a healing mode at the same time. Her main stumbling block was in trying to intellectualize nonlocal healing, just like she'd intellectualized everything she learned in medical school. But gradually both aspects of her activities—the local and the nonlocal, the physical and the mental—came together. Her breakthrough came while she was trying to resuscitate a patient who had suffered a cardiac arrest. While administering cardiopulmonary resuscitation, all of a sudden she found herself in an altered state of awareness. Things were happening effortlessly, perfectly, as if in slow motion. She went into a "zone" like the one many athletes she knew had described. She made the right decisions without trying, as if the decisions were making themselves. Some part of her was functioning superbly outside her rational mind. For a few brief moments she felt pleasantly transformed, and the patient, whom she had expected to die, lived. From that moment on, Pierce understood that the practice of Era III medicine required stepping aside and allowing something to unfold on its own, like that magical moment in childhood when she found that the only way to stay upright on her bicycle was not to try. Letting go was the hardest thing a doctor ever did, but she had stumbled onto the secret. Catherine Pierce had found her groove.

By the time of Rosemary Grey's auto accident, Pierce had begun to think of herself as a healer in addition to being a physician. Whereas being a doctor was an acquired skill, one to which she had dedicated years of her life, it spoke to only one side of her abilities and practice. Healing rounded out her identity. Healing was not something she did, it was a state of being. It was not something she turned on and off, like a defibrillator or a beeper. Healing had become a constant in her life, both inside and outside the emergency room.

As Catherine Pierce stood before Rosemary Grey's unmoving body, she invited it to return to full function. Was there movement? Pierce's heart leaped. After ten years of daily blood and gore, the sight of a body coming to life still moved her so greatly that she could not imagine working anywhere else.

Within five minutes of her arrival, Rosemary Grey's name had been logged into the ER computer as an Era III patient. This triggered a chain of events designed to bring nonlocal healing influences to bear on her from around the world. Chief among them was prayer. One of the cornerstones of Mercy's Era III practice was that protocols were established that functioned instantly and automatically. In the emergency room, response time was critical. One could not wait until a patient needed nonlocal healing interventions before organizing them. That would be like waiting until a patient was hemorrhaging before scouting around for a blood donor. In the ER, systems had to be in place ahead of time.

The *type* of nonlocal interventions selected for the emergency room was important. Pierce knew it might be difficult to marshal widespread support in the community for the Era III program. Although Era III was a concept increasingly accepted by scientists and physicians, it still hadn't trickled down to the community. Yet there was an everyday practice with which the vast majority of people were familiar and that was Era III to the core: intercessory prayer.

Intercessory prayer embodied many of the characteristics of nonlocal events known to scientists. Its power was not diminished by distance, it appeared to function immediately, and it could not be blocked by physical obstructions. But unlike the nonlocal phenomena physicists studied, prayer was fueled by compassion and love. Therefore, although people might balk at "nonlocal healing," Pierce felt they would accept the use of prayer. Prayer had universal appeal. It transcended religion, creed, race, and gender. In building a coalition around nonlocal healing, prayer was the unifying factor Pierce was looking for.

She found an ally in pastoral counselor David Stone, who directed the hospital chaplaincy program at Mercy. Pierce and Stone got on famously from the outset. On their first encounter she explained her goal of bringing nonlocal healing interventions into the ER. Stone listened patiently, then began to chuckle.

"What's so funny?" Pierce asked.

"Your language, the way you beat around the bush. Call it 'nonlocal intentionality' if you like," he said. "I'll stick with 'prayer.'"

"Look. Doctors get uptight about prayer," Pierce explained. "Skeptics say it can't work. If I call it Factor x, there's less backlash."

"Sanitize prayer if you need to," Stone said, "but make sure you don't kill it in the process."

"Don't worry. Prayer's safe. It doesn't need my permission to work."

Every physician called on the services of hospital chaplains from time to time, which meant that Stone could be a valuable ambassador for the Era III project. When physicians heard about prayer in the emergency room, they would ask him what was going on. Stone needed to be able to explain to them why the program made sound, scientific sense. Pierce therefore spent hours discussing the project with Stone. She briefed him on the experiments that had been done in people, animals, plants, and microbes showing the effectiveness of healing intentions. When she saw that Stone was fascinated by the science surrounding

Era III medicine, she breathed a sigh of relief. Not all clerics were enthusiastic about these developments. Some felt that science should keep its hands off prayer. "Five hundred years ago you scientists would have been burned as heretics for 'testing God,'" Stone told Pierce following an extended discussion. How things had changed!

Stone committed himself to the Era III project. Within a month he had met with the city's leading clergy from churches, temples, and mosques. Would they encourage their congregations to participate in a novel prayer effort for emergency room patients? Prayer would be initiated by phone and computer as soon as possible after entry into the ER. The prayer groups must be staffed around the clock, so that no patient would be omitted. Patients would be prayed for by many different religions. Because the prayer effort would be ecumenical, no religion could claim credit if the experiment was successful. Were people willing to pray in concert with people of different religions? Were they willing to put prayer to the test? The patients' clinical course would be compared to ER patients who did not elect prayer. At the end of three years, the results of the experiment would be published in a medical journal. What if prayer didn't work?

Stone's recruiting efforts did not stop at the city limits. He contacted colleagues around the country who shared his interests in healing. One contact led to another. Within six months Stone had cobbled together an international prayer network that included Christians, Jews, Muslims, Buddhists, and Hindus. He discovered that in even the remotest religious outposts there were cybermonks and -nuns linked to the outside world with some of the most sophisticated computer equipment available. Invariably they were grateful to be asked to contribute their prayers. By the time the Era III project was launched, prayer groups were poised to pray from every time zone on the planet. The protocols were set.

Within minutes of Rosemary Grey's arrival in the emergency room, Stone's prayer groups were alerted electronically to her admission. For the first time in her life, Rosemary, unconscious on a stretcher, became the focus of a global prayer effort. This Era III intervention was folded seamlessly with the Era I therapies she was receiving at the time—the monitoring of her vital signs, the electrocardiogram and X rays, the infusion of intravenous fluids, the cleansing of her laceration and abrasions, and the suturing of her wound that was to follow. Pierce knew she was on solid scientific ground in combining these approaches in surgical situations. She was aware of data showing that if patients undergoing cardiac surgery were prayed for, they experienced 50 to 100 percent fewer complications from the surgery than patients who were not prayed for and that their surgical incisions healed faster.

Leslie Milton, the nurse in charge of the emergency room, noticed Grey's movements. Leaning over her patient, Milton gently touched her cheek and whispered, "Welcome back." She took Grey's blood pressure again and reported to Pierce that it was now normal.

When Rosemary was fully awake, Cathy Pierce called to Milton for a suture tray—not vocally, but by thinking it. Milton appeared with the tray immediately. Communicating nonlocally with each other was a private game they played. They often experimented with how long they could function without saying a word. It was time to repair the six-inch laceration in Grey's forehead. As Milton was prepping the wound, Pierce reminded Grey, "You're Era III, but you can do things the old-fashioned way if you like. No need to be a hero." Pierce drew the sheets around the cubicle for privacy and donned the sterile latex gloves Milton handed her.

"What are my options?" Grey asked.

"We can use a little Era I local anesthesia," Pierce elaborated. "Or some Era II hypnosis, imagery, and relaxation; or some Era III intentions and prayer from Milton and me. It's cafeteria style, pick and choose. All or none of the above. Me, I'd have the works. But I'm chicken."

"She's right," Milton added. "The doctor is *always* right."

Rosemary agreed to all three. Pierce ordered Lidacaine, and while Milton upended the vial, Pierce withdrew five cc's into a syringe, changed needles, and began to inject the margins of the jagged laceration. Following the initial sting, Rosemary felt nothing. She had felt strangely peaceful since regaining consciousness, and she sensed her serenity deepening. She knew the two women standing alongside her were responsible. Their casual style was calculated to put her at ease and evoke confidence. Nothing they did was an accident. They even touched her with deliberateness and forethought. Rosemary had lost count of the number of times she had been touched by loving hands—when her blood pressure and pulse were repeatedly taken, her hair brushed aside, her laceration cleansed.

When Cathy Pierce finished the injection, she laid the syringe on the sterile tray. This was a signal to Milton. While they waited for the local anesthesia to take effect, the two women stood silently with closed eyes by Rosemary Grey's stretcher. Grey noticed, and closed her eyes, too, and she drifted into sleep.

Pierce emptied her mind and felt a sense of peace fill her. Then a feeling of oneness enveloped her, her patient, and the nurse at her side. She retraced the path in her mind that led to the universal power and wisdom she believed in, and she asked that this power be lavished on Rosemary Grey. She asked for nothing specific, only that the best outcome prevail. Then she let her feelings of compassion deepen. As she did so a sense of warmth pervaded her, and she knew she was on track.

Pierce was applying a bandage when Rosemary awoke. She had slept through the entire procedure.

"That was nice. What did you do to me?" she asked dreamily.

"We don't *do* things to people around here," Pierce replied. "We try to *be* toward them in a certain way, that's all."

"Since you've been here you've attracted attention from around the world," Milton added. "Your Era III bracelet gave us permission to request healing for you from people everywhere. The requests went out automatically when we logged you into the computer."

"Too bad I had to get myself nearly killed to receive it," Rosemary said. "Can you leave me on your computer indefinitely?"

"For six weeks," Pierce said. "Then you'll be removed automatically. You'll be well by then. Don't worry. The effects of love and compassion don't stop when the intentions cease. They're timeless. Once loved, always loved. No extra charge."

SCIENTIFIC REASONS THAT JUSTIFY THE ADOPTION OF ERA III METHODS

What evidence supports the Era III interventions used on behalf of Rosemary Grey?

1. *Using intentions and prayers to promote healing*

As we have seen, an extensive body of data—approximately 150 studies—supports the ability of thoughts, wishes, and prayers to affect distant biological systems. These experiments involve not only humans but human tissue that has been removed from the body, as well as animals, plants, and microorganisms. Some of these studies are reviewed in chapter 2 and are summarized elsewhere.[24] The fact that these experiments involve animals, plants, microbes, and cells is strong evidence that the effects cannot be attributed to the power of suggestion or belief—the placebo response—as often claimed.

How reliable is the evidence? Studies vary in quality within any area of science, and this field is no exception. As in all fields, one picks the best studies to assess the effects being claimed. Over the past twenty years, researchers in this field have been subjected to such intense criticism by skeptics that the quality of their studies has risen to high levels. British biologist Rupert Sheldrake examined 1,423 papers from different fields of experimental science published in world-famous journals between October 1996 and February 1997. His goal was to establish the prevalence of the use of blind methodologies in experimental research in various scientific fields. In the physical sciences, no blind experiments were found among the 237 papers reviewed. In the biological sciences, there were 7 blind experiments out of 914 (0.8 percent). In the medical sciences 6 of 102 studies (5.9 percent), and in psychology and animal behavior 7 out of 143 experiments (4.9 percent) used such methods. By far the highest proportion of studies employing blind methods—23 of

27 studies (85.2 percent)—was found in parapsychology, the field that explores the effects of distant intentions in healing.[25]

These studies clearly show that *healing intention* is a general term. It can be secular or religious; it may or may not involve prayer. Healing intentions are as varied as the individuals using them. In our culture, with its increasing religious pluralism, this is an important point. Nonlocal healing is often equated with Western-style prayer, but this is only one form it may take. Consider, for example, the study by Sicher, Targ, Moore, and Smith at the University of California, San Francisco, School of Medicine and California Pacific Medical Center, in which "distant healing" proved effective in treating patients with advanced AIDS.[26] The healers in the study described their methods differently. Some called their technique *prayer,* while others favored a secular term such as *healing.* To capture the various methods, the researchers chose the neutral term *distant healing.*

The experiments in distant healing also show that there is no correlation between the effects of prayer and the specific religion of the pray-er. This is an important point for hospitals to consider. Since no single religion can claim superior effects in prayer, hospitals may follow a pluralistic approach such as at the VA Hospital, Duke University Medical Center, in Durham, North Carolina. In recruiting prayer for patients undergoing cardiac procedures, Dr. Mitchell Krucoff, director of the cardiac catheterization laboratory, and his team solicit services from a wide variety of religions worldwide—Protestant, Jewish, and Buddhist. This is a good strategy for institutions because it honors all their constituents. It also prevents any particular religion from claiming success for the prayer, thereby reducing the risk of religious turf battles in the community.

2. *Using intentions to aid wound healing*

In a study involving humans, a practitioner of Therapeutic Touch held her hands a few inches away from surgical wounds and tried to heal them. (Therapeutic Touch is a technique that uses concentrated awareness and focused intention to bring about healing.) In this well-controlled, double-blind study, thirteen of the twenty-three "treatment" subjects were completely healed on day sixteen, while none of the "nontreatment" subjects was healed.[27]

As we saw in chapter 2, in a controlled study in mice, surgical wounds on the animals' backs healed faster and more completely when their cage was held by a healer who directed healing intentions toward them, compared to controls who were merely held.[28]

3. *Using intentions to limit injury to human tissue*

Thirty-two subjects mentally tried to prevent the dissolution of human red blood cells in test tubes containing a weak saline solution. Highly significant differences were found between the "prevent" and control tubes, suggesting that human tissue can indeed be protected from injury by intentions.[29]

4. *Using intentions to sterilize wounds*

Several studies have demonstrated that people can inhibit the growth rates of pathogenic bacteria, fungi, and yeast cells in test tubes under controlled laboratory conditions.[30]

5. *Nonverbal communication between physician and nurse*

A variety of controlled studies and case reports amply document that people can summon the attention of others and convey highly detailed information at great distances, without any intervening sensory processes.[31]

6. *Using intentions to encourage regaining consciousness and awakening from anesthesia*

In twenty-one experiments conducted over a period of several years, healers tried to awaken mice more quickly from general anesthesia. These experiments were increasingly refined. In one variation, only the image of the experimental mouse was projected on a television monitor to the healer in a distant room, who tried to intervene mentally via the image. Nineteen of the twenty-one studies showed highly significant results: earlier recovery from anesthesia in the mice to whom positive mental intent was extended.[32]

Several studies in humans show that one can influence processes related to physical and mental activation, at a distance, when the recipient is unaware of the attempt.[33] This evidence, along with the experiments in mice, suggests that intentions are valuable in nudging people into a state of greater alertness.

7. *"Inviting" a body to heal instead of commanding it to do so*

Several studies have investigated the style or strategy of individuals using healing intentions and prayer. Two techniques have received attention: (a) *directed* methods, in which one seeks to achieve a *specific* outcome, and (b) *nondirected* methods, in which one asks for "the best thing to happen" or "may thy will be done." In experiments with microorganisms, germinating seeds, and human tissue, *both* methods work, compared to controls who do not receive such intentions. The method one chooses depends largely on one's temperament. Introverted, inner-directed individuals usually feel more at home using a nondirected approach, while extroverted individuals tend to prefer a more robust, aggressive, directed method.[34]

C. ERA III CHANGES IN PRACTICE

Rosemary Grey's experience demonstrates the importance of balance—blending the proven methods of *all* the eras of medicine. This point is expressed by internist and researcher David Riley, M.D., medical editor of the journal *Alternative Therapies in Health and Medicine.* "If you don't believe drugs and surgery are important," he says, "try running an emergency room for thirty minutes without them."

Rosemary Grey's story may seem futuristic, but it is already here. Many hospitals are already putting the interventions she encountered into practice, and scores are well on their way.

My wife, Barbara, a nurse-educator and author in the fields of cardiovascular and holistic nursing, and I were recently invited to consult with a large metropolitan hospital about how it might move in these directions. The CEO assembled a meeting of key leaders from the city's business, health-care, and clergy communities to seek their advice and marshal their support. The meeting was essentially an Era III planning session. I presented evidence from controlled, scientific studies, such as those we've reviewed above, that Era III interventions are associated with positive health outcomes. How, I challenged them, can a hospital be justified in not offering these services to its patients? How can they ethically withhold a measure of proven value? How long should they wait before adopting these interventions? Although most of the community leaders were enthusiastic, a few were hesitant. Their concerns centered on spiritual interventions such as distant healing, which they considered too controversial. The spiritual side of Era III could be handled by the clergy, as clergy have always done. Interestingly, no one, including the physicians in attendance, objected to the *evidence* favoring Era III interventions. As the meeting wound down, one prominent woman, who had listened patiently the entire meeting, had the final word. With fire in her voice she said, "If I were really sick and were admitted to your hospital, and you didn't have these approaches available, I'd *really* be angry!" Era III planning meetings such as this are taking place in hospitals across America. Soon, I predict, the above scenario of Rosemary Grey will be coming to a hospital near you.

In the transition to Era III, individual physicians have been leading the way for years. Susan E. Kolb, M.D., is the inspiration for physician Catherine Pierce in the above preview of Era III. Dr. Kolb is a board-certified plastic and reconstructive surgeon in Atlanta, Georgia, and is on the staff of Emory School of Medicine. Plastic surgeons spend a lot of time in emergency rooms, patching up patients who have been damaged by everything from automobiles to chain saws. Dr. Kolb is a not only one of the most technically skilled surgeons I know, she is also a healer. Unlike many physicians with nonlocal healing talent, she doesn't conceal it. She once told me, "Before I inject the local anesthetic and patch somebody up in the emergency room, I draw the curtains around the bed and perform a silent healing ceremony." Also in the operating room, before her patient has awakened from anesthesia, she conveys healing intentions before removing her gloves, mask, and surgical gown. The nurses who work with her are convinced her methods work. On one occasion Dr. Kolb had just finished sewing up the incision and was in the middle of her healing ritual when she was summoned from the operating room. The nurses requested that she come back and "do the other side" so the patient's two halves would heal equally!

Dr. Kolb is only one such physician. I know scores of similar doctors all over the country. Most work quietly in their clinics and hospitals, combining Era III approaches with conventional care. Others are high-profile individuals—deans of medical schools, chiefs of staff, research scientists—all carrying Era III medicine forward.

The inspiration for nurse Leslie Milton in the above story is impossible to specify, because I know hundreds of nurses who model the part. We owe a collective debt to modern nurses and to their predecessors—the healers, midwives, and wise women of ages past who kept alive the flame of healing during the darkest days of the second millennium. Thousands of them gave their lives in the process, as author and researcher Jeanne Achterberg describes in her book, *Woman as Healer.* [35] They nurtured the seeds of Era III medicine, and we are heirs to their courage. This is why most nurses have Era III medicine in their blood, and why the protagonists in the Rosemary Grey account are women.

ADOPTING ERA III: PRACTICAL CONSIDERATIONS

In the hospitals and clinics that are moving in these directions, Era III medicine is never split off from Eras I and II and used in isolation. For example, at San Francisco's California Pacific Medical Center, affiliated with the University of California, San Francisco, School of Medicine, Era III ventures are integrated with Era I and II procedures. Dr. William B. Stewart, chairman of the Department of Ophthalmology, founded the Program in Medicine and Philosophy at California Pacific Medical Center in 1989 to deepen the understanding of health and healing at his institution and to serve as a model for other hospitals. The program evolved into the Institute for Health and Healing (IHH), which is an exemplary Era III venture combining methods drawn from Eras I, II, and III. Rigid protocols to assess staff performance are followed. Peer review, chart review, and case presentations are done weekly at scheduled meetings. The medical center's Committee on Ambulatory Care Quality and Performance Improvement oversees the clinic. Stewart and his colleagues believe that Era II and III approaches must be subjected to the same scientific rigor as conventional methods. Current research projects at IHH deal with the efficacy of remote healing intentions as an adjunct intervention for metastatic breast cancer, the study of Therapeutic Touch and massage in the neonatal ICU, and the in vitro (test-tube) study of the effect of Chinese herbs in combination with standard chemotherapy for breast cancer.[36]

Another groundbreaking program is the Stanford University Hospital's Complementary Medicine Clinic, which opened in April 1998 under the direction of psychiatrist David Spiegel. A variety of Era II and Era III therapies are offered. Among the criteria for selecting the therapies are (1) adequate scientific basis and (2) expertise of a Stanford faculty member to oversee the service or program. In a novel adaptation of an alternative therapy, mindfulness meditation is available for a particularly high-stress group of individuals—patients awaiting organ transplants.[37]

Similar programs exist in every area of the United States. Hospitals wishing to establish programs such as these should keep in mind that there is no formula for doing so. Each institution is unique; each has its individual strengths, weaknesses, and needs. An Era III program must grow organically from within each organization if it is to flourish. Although other Era III programs can be used as models, it is risky to try to copy precisely the program of one institution and graft it onto another.

A common concern of practitioners and health-care institutions is whether or not patients will abandon conventional approaches if they make unconventional methods available. As we've mentioned, this fear seems unfounded. Dr. Wayne Jonas, former director of the National

Center for Complementary and Alternative Medicine at the National Institutes of Health, reports, "More than 80 percent of those who used unconventional practices in 1990 combined these practices with conventional medicine. Patients who use CAM [complementary and alternative medicine] do not harbor antiscientific or anticonventional medicine sentiments, nor do they represent a disproportionate number of the uneducated, poor, seriously ill, or neurotic."[38]

Some health-care professionals are concerned about mixing medicine with spirituality in Era III. They say that hospitals should stay out of the God business, as one critic put it, because religious enthusiasts will proselytize those who are sick and make them feel guilty about becoming sick in the first place. "Physicians shouldn't impose, in any way, spirituality or religion on a patient," warns Stephen G. Post, Ph.D., professor of biomedical ethics at Case Western Reserve University School of Medicine. "The notion of a physician-guru is very worrisome. . . . [A physician] shouldn't go [into religion and spirituality] unless that patient seems receptive."[39]

These are legitimate concerns, and I share them. But hospitals can largely prevent these problems by adopting standards of conduct for Era III healers along the lines of the Code of Ethics of the American Association of Pastoral Counselors. Even though these potential problems exist, hospitals have a responsibility to meet the spiritual needs of their patients if possible, because a sense of spiritual meaning is connected with health outcomes. In a study at the University of Dartmouth Medical School, Dr. Thomas Oxman and his colleagues found that the factor that best predicted survival of coronary bypass surgery and a smooth post-op course was the degree of religious strength and spiritual meaning people found in their lives. This strength must be nurtured in hospitals if patients are to have the best chance of recovery. The majority of patients *want* spiritual support when they are hospitalized. In one survey, three-fourths of hospitalized patients said that their physicians should be concerned about their spiritual welfare, and one-half wanted their doctor to pray not only for them but with them.[40]

III. THE DOCTOR-PATIENT RELATIONSHIP

Era III methods work best when both patient and physician are in agreement about them. When they aren't, serious problems can arise, as in the experience of Denise Dunne of Birmingham, Alabama.

Dunne has done healing work for over ten years. She was involved in a "hopeless" situation with a friend who awoke eight years ago totally blind in one eye. The ophthalmologist determined that the problem was due to histoplasmosis, an infection, and told the man that he would never see again and that nothing could be done. Over the course of a year, however, Dunne began to work with the patient using Reiki, a popular healing technique involving both distant intention and hands-on methods. At the end of a year her friend had regained vision to 20–200, although only in black and white. At the end of two years of Dunne's treatment, her friend had regained color vision and improved to 20–40 with glasses. His visual acuity has remained stable ever since. In spite of his improvement, "his ophthalmologist became *angry* with him because it was 'impossible for him to be healed,' and at the two-year checkup told him, 'Never come back again!'"[41]

What should a patient do when she is open to Era III approaches but the physician is not or when the doctor thinks a patient cannot possibly improve? "Never stay in treatment with a doctor who thinks that you can't get better," advises Dr. Andrew Weil, a proponent of integrative medicine and the author of *Spontaneous Healing.*[42] I agree with Weil but would add that things are looking up. As Era III medicine becomes more widely practiced, instances of "era incompatibility" between physicians and their patients are certain to become less frequent.

The worldviews of doctors and the professors who teach them are undergoing a subtle but profound shift. For example, courses on the role of religious devotion and prayer in healing are currently being taught in approximately fifty U.S. medical schools. This is a historic event, a stunning reversal of the exclusion of these factors from medical education for most of the twentieth century. In addition, conventional medical journals, such as the *Journal of the American Medical Association,* are increasingly willing to publish studies involving unconventional therapies. *JAMA*'s issue of November 11, 1998, was devoted exclusively to the field of alternative medicine. *JAMA* is particularly open to evidence supporting the role of spirituality in health, as evidenced by articles such as "Should Physicians Prescribe Prayer for Health?"; "Getting Religion Seen as Help in Being Well"; and "Religion and Spirituality: Research and Education." New medical journals such as *Alternative Therapies in Health and Medicine* have also arisen, which provide a voice for research that is out of the mainstream.[43] As a result of all these factors, it is becoming easier to find Era III physicians who are open to the principles of Era III medicine.

Of all the areas involved in Era III medicine, the role of spirituality in health is one of the most progressive. I continually find this amazing; until quite recently this was one of medicine's great unmentionables, like sex in the early part of the twentieth century. David Larson, M.D., is president of the National Institute for Healthcare Research (NIHR) in Rockville, Maryland, which conducts and coordinates research on the relationship between spirituality and physical and mental health. Larson believes physicians are ready to accept spirituality into their arsenal of treatment provided that good science backs it up. He emphasizes the dynamism of physicians involved in the day-to-day practice of medicine, and he contrasts this with the inflexibility often seen in academic medicine. "Anyone who thinks [practicing physicians] are stodgy or slow better think again," he states. "The link with spirituality and medicine is amazing. If there's published research, physicians look at it and proceed, whereas in the academic world, it's still seen as a don't-talk variable [an element to avoid discussing]. But," he adds, "that's just like race and gender, where we have [also] been insensitive."[44]

A CURE FOR ALONENESS

Walt Whitman, America's great nineteenth-century poet, attended the wounded as a hospital nurse in the American Civil War. Whitman knew a lot about empathy and compassion for the sick. "I do not ask how the wounded one feels, / I, myself, become the wounded one," he wrote in 1855 in *Leaves of Grass.* His observation sums up the doctor-patient relationship in Era III: a connection in which a doctor and patient become one because of the unity that is implicit in nonlocal mind. All other facets of the relationship between physician and patient that we value—listening, caring, patience, kindness—are included in this single, overarching possibility.

"Being one" with their physician is what patients desperately desire when they are sick and dying, although they do not often say so. The best physicians know this, and they have subtle ways of meeting their patients halfway. For example, when a doctor tells a terminally ill patient, "Things are serious and are not going well. But I will be with you, no matter what," the physician is saying that the patient does not have to face his illness alone and will never be abandoned. When a doctor affirms a commitment in this way, he or she is functioning as an "eternist"—a physician who understands the timeless connections that are implicit in Era III.

A patient of mine was dying from lung cancer, and the day before his death I sat with his wife and children at his bedside. Although he was not a religious person, he revealed to us that recently he had begun to pray.

"What do you pray for?" I inquired.

"I don't pray for anything," he responded in a whisper. "How would I know what to ask for?" I was surprised at his reply. The man was dying; surely he could think of *some* request.

"If prayer is not for asking, what is it *for?*" I pushed him.

"It isn't 'for' anything," he said thoughtfully. "It mainly reminds me I am not alone."[45]

Jiddu Krishnamurti, one of the most revered spiritual teachers of the twentieth century, once asked a small group of listeners what they would say to a close friend who is about to die. Their answers dealt with sympathetic assurances and gestures of compassion. Krishnamurti stopped them short. These comments are indeed helpful, he said, but the one thing that brings most comfort is this: Tell your friend that in his death a part of you dies and goes with him. Wherever he goes, you go also. He will not be alone.[46]

Consider what can happen when a physician dishonors his connection with his patients. Medical ethicist E. J. Cassell relates what happened to his grandmother when she visited a specialist in the 1930s about a melanoma, a skin cancer, on her face. During the course of the visit, she asked the doctor a question, and he slapped her face. "I'll ask the questions here. I'll do the talking," he said.[47]

An admirable movement called patient-centered care has recently arisen within medicine and has attracted a great deal of attention.[48] Patient-centered care puts the patient first by emphasizing the importance of interpersonal factors in healing such as better communication and deeper caring between physicians and patients. Although this is a step in the right direction, I hope that the proponents of patient-centered medicine will extend it even further. I hope they will realize that our connections are not just *inter*personal but *trans*personal as well, that they are nonlocal and limitless. This would bring even greater comfort to patients by allaying their greatest fear: facing illness and death alone.

CHAPTER 6

ETERNITY MEDICINE

And as to you Death, and you bitter hug of mortality,
it is idle to try and alarm me.

WALT WHITMAN
Song of Myself

When I was in medical training and a particularly heroic treatment had been used in a hopeless situation, we young doctors would often joke, "The treatment was a success, but the patient died." Our dark humor concealed what we did not want to acknowledge—that Era I technology will eventually fail and that the dignity of patients can be sacrificed in the process.

In our war on death, Era I's big guns have been turned inadvertently on those they were meant to protect. The result is that great numbers of patients become casualties in the battles for their lives. For example, researchers have found that adverse reactions to drugs kill over 100,000 people a year in U.S. hospitals.[1] This is the equivalent of a passenger jet crashing every day. If this level of death were seen in any other field, it would probably be considered a national scandal. This shows how we have come to accept without complaining some of the darker aspects of Era I medicine.

The metaphor of war crops up repeatedly when we talk about how Era I is used. As Harvard cardiologist Rodney H. Falk says in his article

"The Death of Death with Dignity," "The patient has been relegated to the battleground on which this struggle takes place. Nowhere is this more apparent than in the intensive care units of major teaching hospitals where patients, regardless of mental status or irreversibility of disease, are treated until the bitter end when they lie bloated, bleeding, and crushed after the final flurry of infusions and cardiac massage, death inevitably having caught up." In England, says Falk, this peculiarly American approach to doing everything in life's final moments is viewed with incomprehension and horror. A distinguished senior physician has called this barbaric Era I ritual "mechanical last rites."[2]

Even so, as an internist, I have a deep devotion to Era I medicine. I'm glad it was there when I needed it for my patients, many of whom owe their lives to its efficacy. But I also wish to call it to task for its profound inadequacies and to ask how they might be overcome. Many physicians are asking similar questions. In fact, Era I's shortcomings are often expressed most vividly by doctors themselves, particularly when they become patients and are dying. In "Thoughts of a Dying Physician," Dr. Frederick Stenn, an associate professor of medicine at Northwestern Medical School in Chicago, had this to say on his deathbed:

> The other day, my doctor sat at my bedside just to talk. He assured me my physical complaints will be eased and that he will be in regular attendance. We talked frankly of the dying process and the need of living as I am dying, living to fully appreciate every moment of life. I liked our conversation but it is hard to come by. Most physicians have lost the pearl that was once an intimate part of the medicine, and that is humanism. Machinery, efficiency and precision have driven from the heart warmth, compassion, sympathy and concern for the individual. Medicine is now an icy science; its charm belongs to another age. The dying man can get little comfort from the mechanical doctor.[3]

Why has Era I medicine become such an "icy science"? It is fear of death that drives it on, fear stemming from a thoroughly local view of who we are—isolated minds locked in physical bodies that are drifting toward extermination. Death is the great unmentionable in Era I medicine, the great destroyer, the ultimate indicator that therapy has failed. It was not always this way.

French historian Philippe Aries points out in his book *Western Attitudes Toward Death* that in ancient times humans apparently did not fear death; the person commonly intuited or was forewarned when and how he or she was going to die. This often involved a kind of nonlocal knowing, seeing the future through prophetic visions and other forms of precognitive awareness, which we moderns have come to deny. It also involved accepting death as part of the natural progression of things. When someone faced death, certain rituals were carried out and liturgies were conducted, resulting in the gradual formation of death ceremonies across the ages. Up to the Middle Ages, the dying person knew and presided over the protocol involved in dying. Dying was a public event, and death occurred in the presence of family, children, friends, and neighbors.[4]

Beginning in the Middle Ages, however, the attitude toward death changed from something considered natural, ordinary, and expected to something shameful and forbidden. "Lies began to surround the process of dying," says psychiatrist Patrick B. Friel. "Motivation for lies surrounding death was at first a desire to spare the sick person but rapidly changed to a new sentiment more characteristic of modern times: the topic of death had to be avoided not for the sake of the dying, but for the sake of society and those close to the dying one. The strong emotion caused by what was now the ugliness of death had to be avoided." He observes that around 1930, instead of dying at home, people began going to nursing homes or hospitals. "Dying in the presence of one's family was no longer acceptable. Death took place frequently in a hospital room, and all too often alone."[5]

The change has been astonishing. Today it is difficult to image that a hundred years ago two-thirds of the people who died in the United States were under the age of fifty and that most died in their own beds at home, surrounded by their loved ones and friends. Children saw death continually and considered it part of their reality. Now, two-thirds of dying people are over sixty-five and die in institutions away from family and friends. Children are shielded from death during their formative years, when they have the security and comfort of their parents who might help them deal with it.[6]

In the twentieth century, death became a taboo and replaced sex as a forbidden topic. A mythology has arisen to reinforce the taboo. It is most often applied to children, who are considered most vulnerable to the grotesqueness of death. Formerly, children were told they were brought by the stork, but children were admitted to the farewell scene at the bed of the dying person. Lately we've reversed the fantasies. Today children are taught in their early years about the physiology of love and where babies come from, but when friends or relatives die, they are told bizarre fantasies about what happens to the dying, in an attempt to explain away death.[7]

How do we break the stranglehold of local thinking about illness and death? It involves admitting that we have made colossal errors in defining ourselves only in terms of the physical body. It requires acknowledging our nonlocal side, that part of us that is infinite in space and time and therefore immortal. Yet these steps are not as radical as they might appear, for they are fortified by scientific evidence, as we've seen. This transformation does *not* require abandoning Era I medicine but calls for supplementing, humanizing, and transcending it through Eras II and III.

In the previous chapters I have laid out some of the steps toward this goal. Now I want to focus on certain aspects of Era III that deal specifically with the *timelessness* of the mind. Unlike the temporal-based approaches of Eras I and II, this side of Era III rests not in time but in eternity. I therefore call it Eternity Medicine. Eternity Medicine holds the promise of something that has lately gone missing from the icy science of modern medicine: the promise of surviving death, the assurance of immortality.

We do not have to invent Eternity Medicine out of whole cloth. Throughout history its central message—the nonlocal, infinite nature of consciousness—has revealed itself to humans in a variety of ways. Let's examine some of them.

PRESENCES

There is something about the mind that does not like to be divided. Psychologist C. G. Jung called this a drive toward wholeness, in which all the elements of the self—the various unconscious and conscious aspects of the mind—come together to form a larger entity, the Self. This completion often takes place as people approach death. Therefore Jung called death "a fearful piece of brutality" but also a "joyful event . . . a wedding" through which the psyche "attains . . . its missing half and achieves wholeness."[8]

Our missing half is the nonlocal side of consciousness. The first time many people become aware of it is when they pay attention to the deaths of others. This is why participating in the death of a loved one can leave one feeling joyful and why deathbed rituals sometimes achieve the status of a celebration. Helping us, the survivors, become aware of nonlocal mind is often the departing person's final gift.

The mind's missing nonlocal half sometimes springs to life when someone is dying, by taking the form of *presences.* Presences seem to precipitate from another dimension. They may be sensed as a person with a particular identity, as an ethereal "being of light," an angel, or as an "immaterial something" that is impossible to describe. No matter what form it takes, a presence can bring immense comfort to a dying person. Because presences suggest survival of bodily death and immortality, they are a valued feature of Eternity Medicine.

It's easy to dismiss presences as hallucinations experienced by dying people whose faculties are fading; but when they are also seen by the living, that's another matter. A patient of mine who lay dying in the hospital had talked for days with an angel he perceived standing by his bedside. The angel was dismissed by his wife and daughter as a harmless fantasy. One night around 2 A.M. during their vigil, they became aware that a soft light had begun to fill the room. Unsure of what was happening, the wife turned hesitantly toward the daughter and asked if she saw anything unusual. "It's Dad's angel," the daughter responded. Then both mother and daughter saw an angel move from the light to the bedside, positioning itself at the dying man's head.

A person who taught me much about dying was anthropologist Virginia Hine, who was one of the gentlest, wisest scholars I have ever known. Two days before her husband, Alden, died from cancer, he began to speak completely unintelligibly. When Ginnie leaned over him closely in order to hear better, he kept pushing her away. She left the bedside weeping, afraid he was asking for something that she could not understand. A few minutes later, one of her children with better perception of what was happening came to her and said, "Don't worry, Mom. He wasn't talking to you." Ginnie went back into the bedroom and began to watch and listen with greater understanding and less fear. Alden was still speaking in words she could not understand to a place that seemed to be just above their heads. She realized why he had pushed her away: In hovering over him, she had been blocking his view of something very important.

"Is there someone here?" a family member asked.

Aldie nodded slowly.

"Who is it?"

"Death," Aldie answered. Hine was amazed that when he spoke to the family he was quite understandable but wholly unintelligible when addressing the invisible presence.

"What is death like?" asked Joey, a son-in-law, trying to enter into the experience with Aldie.

"Benevolent."

"Aldie, can you see God?" Joey's question reflected his genuine concern and uncertainty about spiritual matters. Alden, who had long known Joey's doubts, nodded and grinned slightly.

"Where is God?"

"Right behind death."

A few hours later, one of Aldie's daughters was at his bedside and noticed that he seemed to be alternately speaking and listening.

"Is he here again?" she inquired.

"Which one?" Aldie asked.

"Are there two?"

A slow nodding.

"Who is the second one?"

"Love. Go get Mother, I have to give her some."

When Ginnie came into the room and stood beside the bed, he cupped his hands and carefully lifted something and offered it to her. "For just a few moments," she said, "I too was in some other space where all was peace."

From that moment until his death forty-eight hours later, Alden seemed to be in perpetual dialogue with two presences, Death to his left and Love to his right. To his wife, his interaction with the presences seemed to be a sort of conference, as if an important decision were under consideration. Finally he was nodding as if in agreement, and, speaking in words the family could understand, said, "Yes. Yes, I understand. I'm ready." As if falling asleep, he entered a coma and died two hours later.[9]

Virginia Hine memorialized the death of her husband in an inspiring book, *Last Letter to the Pebble People*.[10]

In 1997 researchers Karlis Osis and Erlendur Haraldsson published the results of interviews with almost fifty thousand terminally ill patients and hundreds of physicians and nurses who cared for them. Their findings, which can be found in their book *What They Saw . . . At the Hour of Death,* reveal that Alden Hine's case is a common event.[11]

THE NEARING DEATH EXPERIENCE

Alden Hine's experience is an example of what transpersonal psychologist Kathleen Dowling Singh calls the "nearing death experience." This is not the same as the well-known "near-death experience" popularized by Drs. Raymond Moody, Kenneth Ring, and others, in which dying people sense they are traveling down a tunnel toward a light, experience a sense of completeness and unspeakable joy, encounter helpful beings, and return transformed and committed to a fuller engagement with life. The nearing death experience takes place several weeks to several hours prior to actual death, as Singh describes in her book, *The Grace in Dying: How We Are Transformed Spiritually as We Die.*[12] It is characterized by what Singh calls a "positive depression," a simultaneous withdrawal and brightening of the person who is nearing death, which may be obvious only to those closest to the dying individual. Then the dying person gradually detaches from all but the innermost circle she or he loves most dearly. There is an end to grasping, yearning, and attachment. More time seems to be spent in deep interior spaces. Singh describes this as "a quieting, a hushing . . . communication [that] is essential, deep, often symbolic." She adds, "One begins to get the feeling of entering holy ground when one approaches a person who is nearing death. The quality of the sacred begins to emerge. . . ." One woman told Singh she was "clearing herself out of the way so God could fill her."

The images the dying person sees are archetypal and become clearer as death approaches. They are often symbolic. "One patient told me she felt like an ice cube melting into a glass of water, . . . returning to that which one already is," Singh reports. The images often convey the feeling that many people see their death as "right and fitting and just," Singh finds.

Singh also describes the absorption into wholeness that so impressed Jung—the joining once again with the original source of being. "It is rather profound . . . and encouraging," she states, "to contemplate the possibility that the life and death of a human being is so lovingly calibrated as to automatically produce union with the Ground of Being." This does not mean that people approach the nearing death experience willingly. "My experience is that most people who are dying have absolutely no desire to do it consciously," Singh observes; "they simply are not evolved to the point where that would be a priority. And, yet," she adds, "everyone does seem to enter a transcendence in the nearing death experience. A transformation occurs from the point of terror at the contemplation of the loss of self into the deep, nurturing, ineffable awareness of Unity."[13]

Helping patients through these experiences is part of the practice of Eternity Medicine. This largely amounts to staying out of the way and not interfering with the natural processes of dying. It involves attending to basic needs, alleviating pain, and being available when needed. It means stepping aside in favor of the processes that are bridging naturally to a nonlocal dimension that the dying person now sees.

NONLOCAL KNOWING AND DEATH

Eternity Medicine, because it recognizes the infinite nature of consciousness, regards immortality as a given. This annuls the sting of death, the horror of its finality.

Yet the actual experience of death, even from the Era I perspective, is almost never as grotesque as it is generally regarded. The legendary physician Sir William Osler (1849–1919) studied some five hundred deaths more than a century ago. He concluded that only 18 percent of the dying suffered physical pain and only 2 percent felt any great anxiety. "We speak of death as the king of terrors," he said, "and yet how rarely the act of dying appears to be painful."[14] American physician Lewis Thomas, who was director of research at Sloan-Kettering Cancer Center, notes that in all his years of practicing medicine he witnessed only one agonizing death, in a patient with rabies. Thomas's and Osler's observations affirm the contention of Buddhist scholar Sogyal Rinpoche, author of *The Tibetan Book of Living and Dying:* "We'll all die successfully. Not to worry."[15]

Thomas wondered why death was painless. He came up with an answer based on biological processes and evolution. "Pain is useful for avoidance, for getting away when there's time to get away, but when it is end game and no way back, pain is likely to be turned off, and the mechanisms for this are wonderfully precise and quick. . . . If I had to design an ecosystem in which creatures had to live off each other and in which dying was an indispensable part of living, I could not think of a better way to manage."[16]

I don't doubt that when death is near the brain floods our bloodstream with pain-relieving endorphins and other chemicals that may kick us into a state of painless serenity. But something more is going on, something nonlocal—the sort of thing Annalee R. Oakes, a nurse in Seattle, Washington, encountered when a victim of an auto accident arrived in her emergency room. The woman had nearly died; she had been resuscitated at the accident scene and again while being transported to the local emergency department. On arriving at the hospital, she frantically tried to gain the attention of a police officer investigating the incident. One of the nurses caring for her tried to comfort her but eventually saw that she wanted to communicate with the officer. Since the woman was intubated with a breathing tube in her trachea and could not speak, the nurse summoned the officer, and together they deciphered the message she scribbled on a clipboard. She wrote that her husband was thrown from the car, was in a ditch about a hundred feet from where she was saved, that he was dead and there was no hurry, but that she wanted him found before daylight. The policeman dispatched the message, and rescuers discovered the body where she specified. Her mind had returned to the accident scene and surveyed the details, including not only the location of her dead husband's body but also the precise description of the vehicles involved.[17]

"When the mind and body separate, the mind has ultradimensional qualities unlimited by physical parameters," Oakes states. As an emergency room nurse, she is in a good position to know. "Patients have described their mind's ability to go through walls, follow their body on a cart to the operating room or the intensive care unit and hover outside of the window (building) to watch the resuscitation procedures from a distance. These accounts are validated by details . . . that correctly identify the structures, equipment, names of various items previously unfamiliar to the patients, placement of hardware and personnel within the rooms, clocks with exact time of patient entry to the area and numerous other acts that could not have been known. . . ."[18]

ETERNITY MEDICINE THERAPIES

Practicing Eternity Medicine is disarmingly simple. No formal procedures are required. Eternity Medicine mainly involves conveying empathy, compassion, and love to the sick and dying person. This provides the support for the revelations and wisdom that usually come to the dying—the regaining of their "missing half," which is imperishable. The great task of Eternity Medicine is to facilitate and not obstruct this process.

Schooled as we are in the aggressive methods of Era I, there is a tendency to convert Eternity Medicine from an act of being into an act of doing. This is a problem not only for physicians but for everyone. When patients of mine were dying in the hospital, family members were often nervous about visiting them. They would ask what they should *do* in the dying person's presence. Others would want to know what they should *tell* the dying person. Yet there is something pathetic about people who busily try to instruct others in death. The dying already know, or are in the process of learning, the great lessons. More often it is the dying who teach us, not we them.

Gentle therapies are nonetheless emerging that have a place in Eternity Medicine. They are so subtle and unobtrusive they can hardly be called interventions. Let's look at two of them.

MUSIC

Therese Schroeder-Sheker is a medieval scholar, singer, harpist, and recording artist who made her debut at Carnegie Hall at a young age. She has established a unique program in "music thanatology"—the Chalice of Repose Project—at St. Patrick's Hospital in Missoula, Montana. Through her research in medieval medical practices,

Schroeder-Sheker found that music was used in deathbed rituals at Cluny monastery in France during the Middle Ages. She adapted the actual scores used by the monks and began to play them on the harp at the bedsides of dying patients. In her program at St. Patrick's, she trains harpists to play, sing, and chant the original melodies. Schroeder-Sheker and her team have conducted over two thousand deathbed rituals, and her program has captured national attention. Having witnessed these ceremonies, I can affirm that they are invariably deeply moving, inexpressibly beautiful, and often transformative. They allay pain and fear, and they connect one with the transcendent—Eternity Medicine at its finest.

PRESENCE

The word *presence* is derived from Latin words meaning "to be" and "before" and is related to *essence.* Presence, therefore, embodies our timeless, fundamental nature. When we are genuinely present with someone, we extend to them a helping hand to the infinite—and in that there is healing.

One of the most skillful "practitioners of presence" I ever knew was a feisty coronary care unit nurse who used the term *rail leaning* to describe what she did. One day I asked her what it meant. "If you are actually leaning against a patient's bed rails," she explained, "you naturally want to touch them and give them your undivided attention. Think of it this way," she said with a smile. "Rail leaning is the opposite of the thirty-second U-turn at the foot of the bed, which you docs are so good at!"

The power of presence is delicate and subtle, as author Dawna Markova encountered when she was hospitalized with a serious illness, as described in her book, *No Enemies Within.*

> The one person whose presence I welcomed was a woman who came to sweep the floors with a large push broom. She was the only one who didn't stick things in, take things out, or ask stupid questions.

> For a few minutes each night, this immense Jamaican woman rested her broom against the wall and sank her body into the turquoise plastic chair in my room. All I heard was the sound of her breath in and out, in and out. It was comforting in a strange and simple way. My own breathing calmed. Of the fifty or so people that made contact with me in any given day, she was the only one who wasn't trying to change me.

One night the woman came to Markova's bedside, reached out, and placed her hand on the top of her shoulder, which was one of the few places in her body that did not hurt. Her touch seemed appropriate and natural, and again Markova was comforted by the woman's breathing. It sounded as if she were saying two words with each breath, one on the inhale, one on the exhale: As . . . Is . . . As . . . Is . . .

On her next visit, the woman looked penetratingly at Markova with great kindness and said simply, "You're more than the sickness in that body." Markova was too sedated with pain medication to ask questions. But throughout the following day she kept mumbling those words to herself: "I'm more than the sickness in this body. I'm more than the suffering in this body." She recalls, "I remember her voice clearly. It was rich, deep, full, like maple syrup in the spring. I reached out for her hand. It was cool and dry. I knew she wouldn't let go." Then the woman continued, "You're not the fear in that body. You're more than that fear. Float on it. Float above it. You're more than that pain." Markova began to breathe deeper, as she did as a child when she wanted to float in a lake. She remembered floating in Lake George as a five-year-old, floating in the Atlantic Ocean at Coney Island at age seven, and in the Indian Ocean off the coast of Africa at age twenty-eight. "Without any instruction from me," Markova states, "this Jamaican guide had led me to a source of comfort that was wider and deeper than pain or fear."

By any measure, the Jamaican janitor was an Era III healer. She helped Markova expand beyond the confines of her body and her pain, to contact wider dimensions of the self where suffering was minimized.

By her mere presence and the sounds of her breath she set healing in motion. Her presence lingers. "It's been fifteen years since I've seen the woman with the broom," Markova says. "I've never been able to find her. No one could remember her name; but she touched my soul with her compassionate presence and her fingerprints are there still."[19]

Cicely Saunders, who founded the modern hospice movement, understood deeply the power of presence to give comfort. The needs of dying patients, she said, can be summed up in three simple words: "Watch with me." This does not mean, Saunders states, "Take away my pain and fear" or even "Help me understand what is happening," but merely "Be there with me."[20]

Saunders's insight has been taken to heart by Cathie Guzzetta, R.N. PH.D., a leader in the field of cardiovascular and holistic nursing. Guzzetta is a nursing research consultant at Parkland Memorial Hospital and Children's Medical Center in Dallas, Texas. She has long been convinced that presence is a vital factor in healing. By *presence* she does not mean physical nearness but an immaterial factor that encourages the flow of empathy and compassion and therefore healing. To bring more presence to bear, she developed a proposal that challenged the widespread practice of prohibiting family members from attending the cardiopulmonary resuscitation (CPR) procedures being carried out on their loved one. According to her plan, family members, if they so chose, would be escorted into the room by a nurse, who would remain at their side and explain to them every step of the CPR process. Although preliminary experience at other institutions had already revealed that this brings immense comfort to families, regardless of whether or not the resuscitation attempt is successful, the staff physicians at Parkland vigorously objected. They feared that if family members were allowed to witness this grim ritual they would become hysterical and interfere, and they might sue the physicians if the resuscitation attempt were not successful. Would they try Guzzetta's bold plan? "Not 'til pigs fly!" thundered one staff physician. However, with the support of Dr. Ron Anderson, Parkland's CEO and a genuinely humanistic physician,

Guzzetta gained final approval after a five-year struggle. Since the program was initiated, many of the staff physicians who originally opposed it have become firm supporters. And one of Guzzetta's nursing colleagues has hung a poster in her office depicting three pigs with wings.

Family presence during resuscitation brings a nonlocal component to CPR, which is perhaps the most mechanical intervention ever devised in the annals of Era I medicine. As family members observe the attempt to save the life of their loved one, they are extending their intentions, prayers, and wishes to the patient and also to the physicians involved. No matter what the CPR outcome proves to be, a bonding often takes place between the family members and the medical and nursing team—an empathic, nonlocal connection.

As an aside, I am gratified that this development involves Parkland, which is the main teaching hospital of Southwestern Medical School, where I trained. I cannot think of Parkland without recalling a shocking encounter with death I had there as a first-year medical student.

On a beautiful, sunny day in November 1963, I returned to Parkland from lunch. As I entered the rear of the hospital via the emergency room, something didn't seem right. There were too many security officers, policemen, and ambulances, even for Parkland. Just inside the emergency entrance were three pay phones. To my surprise, I saw that one of them had been commandeered by the national television reporter Robert Pierpoint, who I recognized immediately. Seeing my short white coat and stethoscope, Pierpoint identified me as a medical student and blurted, "Can you help me?" Without waiting for a reply he said tersely, "The president has been shot. He's inside. Please hang onto this phone and guard it with your life." I took the phone while Pierpoint dashed down the hall to the trauma room where President John F. Kennedy lay. Pierpoint shuttled back and forth between the trauma room and the telephone for more than an hour, rattling off briefings to his New York City news bureau. While he was away from the phone, I provided descriptions to the New York office of the bedlam I was witnessing. When it was certain that the president was dead,

Pierpoint thanked me and moved on with the presidential entourage. By this time I was emotionally numb. I joined three of my fellow medical students who had also been caught up in the event, and we left Parkland for a bar, where we sat for hours in stony silence.

In helping humanize the process of dying, Parkland is pointing the way to Eternity Medicine. I pray that this will help dispel the lingering sense of tragedy that has hung like a shroud over Parkland and Dallas since those dark moments.

Eternity Medicine requires us to consider three controversial areas connected with illness and death: near-death experiences, visitations from the dead, and reincarnation. For too long these events have been dismissed as archaic fantasies for which there is no evidence. In the past three decades, however, they have attracted the attention of serious scholars. These phenomena are connected with illness and dying, and they point like an arrow to a timeless quality of the mind. For these reasons they belong to Eternity Medicine. Let's take a closer look at them.

NEAR-DEATH EXPERIENCES

One way of experiencing the lessons of Eternity Medicine is to die—nearly.

In 1975 psychiatrist Raymond Moody published a landmark book, *Life After Life,* which had a stunning effect on the Western understanding of death. Moody reported patients who, having nearly died, returned with memories of traveling through a tunnel of light, reliving all the events of their life, and meeting gentle, wise beings in another dimension. After recovery they felt transformed and inspired to live a life of service to others. Moody named the phenomenon "near-death experience," or NDE. It was as if a dam had burst. Following the publication of *Life After Life,* a flood of people came forward to reveal their own stories. Some polls estimate that currently there are thirteen million Americans, roughly 5 percent, or one in twenty, who have undergone a near-death experience.[21]

August L. Reader, M.D., a clinical associate professor of ophthalmology at the University of Southern California School of Medicine, Los Angeles, experienced just such an event. In his own words:

> At approximately 1:30 A.M., I experienced the sudden onset of severe, crushing chest pain over the left side of the chest, which quickly radiated down my left arm and up into my left neck. . . . Then my heart suddenly stopped beating. A panic rose in me. . . . As soon as I had the thought of letting go and releasing myself to the arms of God, there was a sudden release of the pain and the fear and a complete sense of bliss and understanding. And in that moment of understanding, there was a panorama . . . of everything that occurred in my life, from the most trivial detail to the most important events, all displayed equally and with no favor, connected by golden threads, showing me that everything that had occurred to me in my life was important, was part of who I was, and was essential to who I had been. And in that knowledge was the understanding that my life had been worthwhile and I had nothing to regret in dying. [Then] . . . I was being projected down a long, rocky tunnel of grayish brown, at the end of which was a bright light.
>
> . . . As I looked, I [saw] . . . faces . . . recognizable as everyone I had known, both alive and dead. Standing at the entrance to this expanse was my father, who had been dead for seven years . . .
>
> I awoke to the alarm clock the next morning, in my bed with the television still on, looking around me and realizing that I had not died but was still in the world of the living. At that point I knew that something very significant had happened. I immediately went to my doctor for physical check-up, was subjected to an ECG [electrocardiogram], cardiac enzymes, and a treadmill test, all of which were normal.[22]

Reader had suffered a transient irregular heartbeat, which normalized spontaneously. He narrowly avoided a heart attack and sudden death. As a neuroophthalmologist trained in the workings of the brain and visual systems, he felt compelled to investigate his NDE scientifically and rationally. He knew that his heart had stopped pumping effectively and that the blood flow to his brain was diminished. Did his oxygen- and blood-starved brain produce the experience? His analysis, published in the journal *Alternative Therapies,* is one of the most profound examinations of the physical side of NDEs ever written.

Yet a physical analysis of the experience, he discovered, raised more questions than it answered. For example, why should someone whose brain is deprived of blood and oxygen experience an overwhelming sense of unity with all things? Why the desire to embrace God and all of humanity? Why should someone, on recovering, put the welfare of their community above self? And why, he asked, is this experience accompanied by such a profound feeling of Spirit? "The only answer that I can give at present is the one that I have received from the experience itself, and is the same answer given by all known religious masters. The answer comes in a love that is so profound, deep, and unifying that it seems that it can only come from a Universal Presence, and from nowhere else."

Reader suggests that the ultimate significance of the near-death experience lies outside the brain, not in it. After years of researching his experience he concludes,

> I have explored many of these paths, from Kabalism to Buddhism, from shamanism to yoga. . . . The final common path is the same. In the end, we each have to face that moment of transition to the next plane of existence. . . . The more we know of it, the easier it is to cope with the everyday stresses of life. In that is healing—spiritually, psychologically, and physically—for the individual. For once you learn how to die, you may then be able to fully live.

Like Reader, millions of individuals have experienced through a near-death experience the wholeness and unity, bliss and perfection, that underlie Eternity Medicine. Many of them are so inspired to live a life of service following their recovery that they practically become "eternists," practitioners of Era III medicine. An example is author

Dannion Brinkley, who, following *two* near-death experiences, underwent a transformation so profound that he dedicated his life to serving dying people in hospices. Brinkley's efforts have become widely known. He has recruited so many volunteers that he has single-handedly reinvigorated the hospice movement throughout the United States.[23]

VISITATIONS FROM THE DEAD

Because most health-care professionals have been trained in Era I medicine, they generally dismiss visitations from the dead as signs of mental pathology. In contrast, Eternity Medicine is cordial to this possibility. By viewing the mind as infinite in time, it grants visiting rights to the deceased.

In 1995, Miami attorney Arthur S. Berger investigated what he called "The Will Case." It involved James L. Chaffin, a North Carolina farmer who made a will in which he left his farm to one son and nothing at all to his wife and other three sons. Fourteen years later he tried to rectify matters by writing out a new will and signing it, then inserting it in an old Bible. Chaffin never told anyone he had done so. He died, and the original will was probated. Then one night one of the sons who had been left out of the first will saw his deceased father standing at his bedside. His father told him that he would find the new will in the father's old overcoat pocket. The son found the overcoat, along with a note written by old Chaffin stitched in its inside pocket. It contained a reference to the chapter in the Bible where the second will could be found. The lawyer in the case, despite all his skill, could find no reason to suspect the sincerity and honesty of the parties involved, and the second will was probated in court exactly as Chaffin had written it.[24]

Dream visitations sometimes convey healing. Meredith, a forty-three-year-old Saskatchewan homemaker, tripped on the stairs one day and sustained a painful injury to her ankle. Her doctor X-rayed her, found a hairline fracture, gave her crutches, and told her to stay off her foot for six to eight weeks. The next night she dreamed she was in a beautiful, peaceful setting, and her sweetheart Vic, who had died from cancer seven years earlier, appeared. Vic said he wanted to take a look at her foot. "I don't think you can do anything about it. You're not a doctor," Meredith protested, as if he were real. Vic took her foot in his hands and moved it, and Meredith heard a click. Then the pain disappeared. Vic told her she would be able to walk on awakening but that she should treat the ankle gently and give it time to heal. She awoke immediately, got up, and without thinking bore her full weight on the ankle. The pain had disappeared, as in the dream.[25]

In spite of the fact that the Era I perspective says they can't come back, the deceased haven't got the message. They are reappearing in droves. When bereavement researcher W. D. Rees interviewed 227 widows and sixty-six widowers, he found that 46.7 percent of the 293 people had visitation experiences, divided equally among the sexes. A sense of the presence of the deceased spouse occurred in 39.2 percent of those interviewed, visual experiences 13.3 percent of the time, auditory experiences 13.3 percent, and spoken messages from the dead individual 11.6 percent of the time.[26] Other investigators have confirmed these findings. They involve both sexes, cut across every culture, are common in the largest cities and smallest villages, take place in the first days of grief and years later, and happen to atheists, agnostics, and believers alike.

Sometimes the visitation involves a feeling or a kind of knowing. A therapist who conducted a support group for widows reported comments such as, "I'll be sitting at the desk, and suddenly I know he's in the room. I can feel his presence." Or, "I know I wasn't asleep, because suddenly I woke up and could feel him sitting on the bed. I could smell his after-shave lotion. I touched his hand and said, 'Your hand isn't cold, but I know you're dead.' He just smiled. . . ." And, "I know you'll think I'm crazy, but twice before I went to sleep I asked Barney to help me find some papers I had misplaced, and in the morning I knew exactly where to look."[27]

Anyone who counsels bereaved individuals has an opportunity to practice Eternity Medicine. Bereaved individuals are in serious need of the healing comfort offered by Eternity Medicine. In almost every study done in the Western world, the mortality of the surviving spouse during the first year of bereavement has been found to be two to twelve times that of married persons the same age. In the United States, around seven hundred thousand people aged fifty or more lose their spouses annually. Of these, thirty-five thousand die during the first year of grief. Researcher Steven Schleifer of New York's Mount Sinai Hospital calculates that 20 percent, or seven thousand, of these deaths are directly caused by the loss of the spouse.[28] The evidence suggests that it is the ongoing, unrelieved despair of grief and bereavement—the giving up—that causes the body to give up as well. The Era I view intensifies and prolongs grief by emphasizing that physical death is the end of everything, including consciousness. In contrast, Era III emphasizes immortality, which is why Era III is much more effective in helping people navigate one of the most vexing events they will ever experience.

Yet terms such as bereavement *counseling* do not do justice to Eternity Medicine. *Counseling* is a term that is more appropriate to Era II, in which some type of psychotherapy is applied by one individual to another in hopes of bolstering the self-esteem and ego strength of the client. In Eternity Medicine, we are seeking not to strengthen the sense of self, but to help an individual *transcend* it by connecting with a nonlocal dimension of being.

THE "ETERNIST"

During my years as an internist, there were many times when I practiced Eternity Medicine without realizing it. One instance involved a couple who came to my office to discuss the impending death of Amy, their six-year-old daughter, who suffered from leukemia. This is one of the greatest challenges physicians ever face. What can one possibly say to bring comfort and hope to heartbroken parents whose child is dying? The parents wanted to know if I believed her treatments were adequate. Was enough being done? After discussing these questions we focused on the pain and grief these loving parents felt, and tears flowed all around. When it came time for them to go, on an impulse I picked up a book I had just finished reading. "This may give you a different way of looking at Amy's death," I said. "When you've read it, let's talk again." The book was *Children Who Remember Previous Lives: A Question of Reincarnation,* by Ian Stevenson, M.D., Carlson Professor of Psychiatry and director of the Division of Personality Studies at the Health Sciences Center, University of Virginia. It discusses cases of young children who insist that they have lived before and who have memories, often highly detailed, that support this possibility. Stevenson's fanatically precise investigations offer solid hope that death is not final. They help people understand that reincarnation is not as outrageous as we've been taught—that, as Voltaire put it, "It is not more surprising to be born twice than once."[29] The next week the couple returned. Although Amy's medical condition had worsened, her parents had a serenity and strength not present the week earlier. "Thank you for the book," the mother said. "The children's stories brought our religious beliefs down to earth and made our hopes real."

I then began to offer Stevenson's book to my adult patients who faced terminal situations and to their loved ones as well. The result was always the same: greater comfort and hope. Not one individual objected to the idea of reincarnation on religious grounds. In addition to being an internist, I was being an "eternist"—a physician who recognizes our timeless nature and helps a patient understand this reality. This, to me, is the ultimate skill one can practice as an Era III healer.

As an Era III physician, however, I continue to honor Eras I and II because of their important contributions to human welfare. I'm glad that when one becomes an eternist, one doesn't have to stop being an internist. As a doctor, I've needed *both* internal and eternal medicine. So, too, have most of my patients.

An incident in the life of Mahatma Gandhi shows that each era of medicine does not eradicate, but expands the stages that came before it and that one can pick and choose therapies from different eras when confronted with a particular problem. In January 1924 Gandhi was in jail in India for leading his country's nonviolent struggle for independence from Great Britain. Gandhi advocated the exclusive use of natural approaches to health, such as exercise, fresh air, vegetarianism, and prayer. His views had no place for Western-style interventions, such as drugs and surgical procedures. On the evening of January 11, however, he became ill with acute appendicitis and was transferred from jail to Sassoon Hospital in Poona. There he was met by Colonel Maddock, the British surgeon general at the hospital. Maddock was respectful of Gandhi's views, but he was certain that surgery was called for. He sent urgent messages to two Indian doctors who were close friends of Gandhi, in hopes that they could perform the surgery, but neither was available. Twenty-four hours later it was obvious that if Gandhi's life was to be saved, surgery was necessary and speed was essential. Gandhi drew up a statement expressing his fullest confidence in Colonel Maddock. He wanted his followers to know that the operation was his choice and had not been forced on him by the British. A few minutes later the operation began. Gandhi's appendix was successfully removed, his life was saved, and he was able to lead one of the most remarkable struggles for freedom the world has ever known.[30]

If the great Gandhi, whose saintly vision was fixed on the eternal, could find a place for multiple approaches to healing, so can we. Like Gandhi, we should keep our options open. If the saints need surgery, so may we. So let us bow deeply in gratitude to Eras I and II for their majestic contributions. And as we venture into Era III, let us prepare for the transformation not only of our bodies but of our hearts as well.

Is *transformation* too lofty a word to apply to a system of healing? Perhaps to Eras I and II, but not to Era III, because Era III has the capacity not only to heal our bodies but also to transfigure our sense of our origin, nature, and destiny. The key to this transformation is our awareness of the nonlocal nature of our own mind—that it is infinite, indestructible, and immortal.

I have often pondered why it is medicine through which our culture is confronting these great questions once again. Why not just through our religions, which long have spoken of these issues? Medicine's current involvement in life's greatest mysteries is no accident. Healing has always been the common meeting ground between the physical and the spiritual. When we experience illness, even something as minor as a cold or the flu, we are brought face-to-face with our vulnerability, and if we look ahead, which we are prone to do at such moments, we can see death. As a result, illness has always been a natural breeding ground for spiritual concerns. There is nothing obscure about this. All doctors, if they listen to what their patients tell them, come to realize the transformative power of illness. I recall, for example, many patients who made comments such as "My heart attack was the best thing that ever happened to me" or "I wish I had developed cancer twenty years earlier." Their illnesses were opportunities for seeing deeper, for going further. Serving people who are undergoing these life-changing events is one reason why medicine has always been considered a priestly function and why becoming a physician has always been regarded as a spiritual path.

But something new and unprecedented is pushing medicine to center stage. As we have seen, for the first time in human history, science—data, fact, evidence—is affirming our nonlocal nature and our imperishable Self. This development adds immeasurably to the transformative power of Era III medicine. As a result, ours is a landmark period in history. Never before have humans enjoyed such an advantage.

We should be careful how we respond to this opportunity. This chance may not come our way again.

POSTSCRIPT

For more than a century the profession of medicine has tried to become increasingly scientific and technical, because this is where we believed the future of healing lay. Now a monumental shift is occurring, set in motion not by physicians but by society at large.

Currently almost half of adult Americans visit a practitioner of alternative medicine yearly.[1] This exceeds the total number of visits to frontline physicians such as family physicians, internists, pediatricians, and gynecologists. This is one of the most remarkable social movements occurring in the twentieth century, and orthodox physicians have been baffled about its causes. Finally they are beginning to understand why. As mentioned, in 1998, Dr. John A. Astin of Stanford University School of Medicine reported the findings of a national survey in the *Journal of the American Medical Association* in an article entitled "Why Patients Use Alternative Medicine." "Users of alternative health care are more likely to report having had a transformational experience that changed the way they saw the world," he stated. "They find in [alternative therapies] an acknowledgment of the importance of treating illness within a larger context of spirituality and life meaning. The use of alternative care is part of a broader value orientation and set of cultural beliefs, one that embraces a holistic, spiritual orientation to life."[2] In other words, technological medicine is not enough. People want their medical care grounded in spirituality.

I have described the shift toward spirituality by focusing on three periods of development within modern medicine, Eras I, II, and III. This approach allows us to tease out the mechanical, the mind-body, and the spiritual elements of modern medicine. It permits us to see how one development gives rise to another and how they are compatible with one another. This helps explain why millions of people believe there is no contradiction among the various physical, mind-body, and spiritual approaches and why they use them together.

Some physicians are having greater difficulty implementing these concepts than their patients. They find the science underlying Era III to be challenging, because the basic idea of nonlocal mind is foreign to the science in which they were trained. Moreover, some forces are trying to steer medicine back to Era I, to make healing a totally mechanical, objective affair. This effort is misconceived, because healing has never been as objective and rational as the mechanical enthusiasts believe. There never was a golden age of mechanical medicine, so there is nothing to which we can return.

I am confident, however, that physicians will eventually embrace Era III because of two main reasons. Era III rests in science, which is the language physicians understand, and the scientific evidence favoring Era III is steadily increasing. In addition, the path of the physician since antiquity has been considered a spiritual path and remains so. This suggests that most physicians possess an intrinsic sensitivity to spiritual issues, which further enables them to respond to Era III.

Of interest to physicians, psychologist and author Charles T. Tart, Professor Emeritus, University of California at Davis, has created an online journal devoted to nonlocal, transcendent experiences that physicians and scientists have reported. It lets them express these experiences in a safe place, and debunks the sterotype that "real" scientists don't have "spiritual" or "mystical" or "psychic" experiences. The online site is called *The Archives of Scientists' Transcendental Experiences* (TASTE). The internet addresses for TASTE are <http://psychology.ucdavis.edu/tart/taste> or <http://www.issc-taste.org>.

In making a place for Era III, we—physicians and laypeople alike—must handle it with care. The greatest danger is the temptation to use nonlocal mind, the foundation of Era III, as the latest superdrug or surgical breakthrough. If we do this we shall lose the spiritual component of Era III, and it will become just another aspirin. Era III is practical, as we have seen, but it is more—a signpost pointing toward immortality.

I am aware that the transitions I have described offer great challenges to physicians, who are already challenged on every hand. In fact, this is not a happy time in medicine. I do not recall the last time I had an upbeat conversation with a practicing physician. There is the perpetual scowl, the hundred-yard stare at the end of the day, the lament that "things aren't like they used to be" and are getting worse. Chronic depression is epidemic among doctors, and one wonders whether the profession as a whole is not a candidate for Prozac or at least a bit of talk therapy. Thousands of physicians are looking for a way out. They go about their daily rounds with one eye trained on their practice and the other peeled for an escape hatch into another line of work. But if medicine is in crisis, perhaps it is a good thing.

Yellowstone National Park is one of the gems of the planet. My wife and I have made pilgrimages to it for years, drawing strength and inspiration from its canyons, meadows, streams, and wildlife. When the park was ravaged by fire in 1988, we were disconsolate, but only temporarily. Soon it became apparent that the fire was not a cremation but a purification. We were astonished to see the profusion of life that quickly arose. It is as if Yellowstone has shaken off the cobwebs and is reinventing itself. New trees are now growing among the blackened skeletons of the old forest, and there are burgeoning populations of wildflowers, birds, bison, bears, and elk—all made possible by the enriched habitat resulting from the fires.

In a sense, medicine is burning, as old ideas and methods are fading on every hand. But medicine's fires are purifying; new life is emerging from the ashes, as it always does. The reinventors are stepping forward, and healing is in the wind. The rebirth has begun.

NOTES

INTRODUCTION

1. On Hippocrates, see L. Bonuzzi, "About the Origins of the Scientific Study of Sleep and Dreaming," in *Experimental Study of Human Sleep: Methodological Problems,* ed. G. Lairy and P. Salzarulo (Amsterdam: Elsevier Scientific, 1975), 192. On Plato and dreams, see H. McCurdy, "The History of Dream Theory," *Psychological Review* 53 (1946): 225–33. On Galen, see Robert L. Van de Castle, *Our Dreaming Mind* (New York: Ballantine Books, 1994), 64–65. Cicero's "Argument Against Taking Dreams Seriously" is found in *The World of Dreams,* ed. R. L. Woods (New York: Random House, 1947), 204.

2. I. Regardie, *The Philosopher's Stone* (St. Paul: Llewellyn Publications, 1970), 90, quoted in Randolph Severson, "The Alchemy of Dreamwork: Reflections on Freud and the Alchemical Tradition," *Dragonflies* (Spring 1979): 109.

3. Asclepius, *A Holy Book of Hermes Trismegistus,* addressed to Asclepius, in *Hermetica,* ed. and trans. Walter Scott (Boulder: Hermes House, 1982), 295–96.

4. Libellus X, *A Discourse of Hermes Trismegistus, The Key,* ed. and trans. Walter Scott (Boulder: Hermes House, 1982), 203–5.

5. Brian Inglis, *Natural and Supernatural: A History of the Paranormal* (Bridport, Dorset, England: Prism Press, 1992), 34.

6. Josiah Gregg's story is found in David Lavender, *Bent's Fort* (Lincoln: Univ. of Nebraska Press, 1954), 125. Victorio and Lozen's story is told by James Kaywaykla, quoted by Eve Ball, *In the Days of Victorio: Recollections of a Warm Springs Apache* (Tucson: Univ. of Arizona Press, 1992), 11.

7. Samuel Taylor Coleridge, quoted in Michael Grosso, *Soulmaking* (Charlottesville, VA: Hampton Roads, 1997), 181.

8. Larry Dossey, *Recovering the Soul* (New York: Bantam, 1989), 1–11.

9. The quotations from Thomas Kuhn and Stephen Jay Gould in this section are from T. Peter Park, "Too Many Anomalies, Not Enough Time," *The Anomalist* (1997): 4–7.

CHAPTER 1

1. George Washington, *Last Will and Testament,* from John C. Fitzpatrick, ed., *Writings of George Washington from the Original Manuscript Sources, 1745–1799,* 39 vols. (Washington, DC: Government Printing Office, 1931–1944), 37, 275–305.

2. Tobias Lear, *Letters and Recollections* (New York: Macmillan, 1906), 129–41; Willard Sterne Randall, *George Washington: A Life* (New York: Henry Holt, 1997), 500–502.

3. Willard Sterne Randall provides compelling descriptions of Washington's illnesses throughout *George Washington: A Life.*

4. John Polkinghorne, *Science and Providence: God's Interaction with the World* (Boston: Shambhala/New Science Library, 1989), 75.

5. On the mortality rate in ancient Rome and North Africa, see John Cairns, *Matters of Life and Death* (Princeton: Princeton Univ. Press, 1997), 20. Nancy Tomes offers infant mortality rates for nineteenth-

century New York in *The Gospel of Germs: Men, Women, and the Microbe in American Life* (Cambridge: Harvard Univ. Press, 1998), 33. On semirural districts of England in 1860, see Cairns, *Matters of Life and Death,* 20.

6. Joel E. Cohen, "The Bright Side of the Plague," *The New York Review,* March 4, 1999, 26–28.

7. Tomes, *Gospel of Germs,* 26–27.

8. Cairns, *Matters of Life and Death,* 3.

9. Cairns, *Matters of Life and Death,* 31–33.

10. Cairns, *Matters of Life and Death,* 35, 38.

11. Cairns, *Matters of Life and Death,* 35–36.

12. Carl Sagan, *The Dragons of Eden* (New York: Random House, 1977), 7.

13. John A. Astin, "Why Patients Use Alternative Medicine," *Journal of the American Medical Association* 279, no. 19 (1998): 1548–53.

14. Michael Grosso, *Soulmaking* (Charlottesville, VA: Hampton Roads, 1997), 85.

15. C. G. Jung, *Psychology and the East,* trans. R. F. C. Hull (Princeton: Princeton Univ. Press, 1978), 126–27, 131, 214.

16. Grosso, *Soulmaking,* 85.

17. John Searle, *Journal of Consciousness Studies* 2, no. 1 (1995): front cover quotation; Jerry A. Fodor, "The Big Idea," *The Times Literary Supplement,* July 3, 1992.

18. Grosso, *Soulmaking,* 200.

19. Lama Govinda, "The World View of a Mahayana Buddhist," a conversation with Lama Govinda conducted by Emily Sellon and Renée Weber, *ReVision* (Summer/Fall 1979): 28–42.

20. Gertrude Stein, *The Autobiography of Alice B. Toklas* (1933), from *Bartlett's Quotations,* sixteenth edition, Justin Kaplan, general editor (Boston: Little, Brown and Company: 1992), 627.

CHAPTER 2

1. Kimberly A. Sherrill and David B. Larson, "The Anti-Tenure Factor in Religious Research in Clinical Epidemiology and Aging," in *Religion in Aging and Health,* ed. Jeffrey S. Levin (Thousand Oaks, CA: Sage Publications, 1994), 149–77.

2. Ian Wilson, *The After Death Experience: The Physics of the Non-physical* (New York: William Morrow, 1987), 54–58.

3. Wilson, *After Death Experience,* 57–58.

4. Edgar Mitchell, *The Way of the Explorer,* with Dwight Williams (New York: G. P. Putnam's Sons, 1996), 51.

5. Lawrence LeShan, *The Medium, the Mystic, and the Physicist* (New York: Viking, 1974).

6. Lyall Watson, *The Nature of Things: The Secret Life of Inanimate Objects* (Rochester, VT: Destiny Books, 1992).

7. Bernard R. Grad, "Some Biological Effects of Laying-On of Hands: A Review of Experiments with Animals and Plants," *Journal of the American Society for Psychical Research* 59a (1965): 95–127.

8. Grad, "Some Biological Effects."

9. Grad, "Some Biological Effects."

10. Grad, "Some Biological Effects."

11. B. Grad, R. J. Cadoret, and G. I. Paul, "The Influence of an Unorthodox Method of Treatment of Wound Healing in Mice," *International Journal of Parapsychology* 3 (1961): 5–24.

12. Grad, "Some Biological Effects," 124–25.

13. Fred Sicher et al., "A Randomized Double-Blind Study of the Effect of Distant Healing in a Population with Advanced AIDS: Report of a Small-Scale Study," *Western Journal of Medicine* 169, no. 6 (1998): 356–63.

14. David Van Biema, "A Test of the Healing Power of Prayer," *Time,* October 12, 1998, 72–73.

15. Bonnie Horrigan, "The MANTRA Study Project," interview with Mitchell W. Krucoff, *Alternative Therapies in Health and Medicine* 5, no. 3 (1999): 75–82.

16. Seán O'Laoire, "An Experimental Study of the Effects of Distant, Intercessory Prayer on Self-Esteem, Anxiety, and Depression," *Alternative Therapies* 3, no. 6 (1997): 39–53.

17. Kenneth S. Cohen, *The Way of Qi Gong* (New York: Ballantine, 1997).

18. David J. Muehsam et al., "Effects of Qigong on Cell-Free Myosin Phosphorylation: Preliminary Experiments," *Subtle Energies* 5, no. 1 (1994): 93–108.

19. Garret L. Yount, Yifang Qian, and Helene S. Smith, "Cell Biology Meets Qi Gong" (paper presented at the annual meeting of the Society for Scientific Exploration, Las Vegas, NV, June 5–8, 1997). Abstract of presentation in *The Explorer: Newsletter of the Society for Scientific Exploration* 13, nos. 2, 3 (1997): 1.

20. William G. Braud, "Distant Mental Influence of Rate of Hemolysis of Human Red Blood Cells," *Journal of the American Society for Psychical Research* 84, no. 1 (1990): 1–24.

21. Randolph C. Byrd, "Positive Therapeutic Effects of Intercessory Prayer in a Coronary Care Unit Population," *Southern Medical Journal* 81, no. 7 (1988): 826–29.

22. Howard Wolinsky, "Prayers Do Aid Sick, Study Finds," *Chicago Sun-Times,* January 26, 1986, 30. The following description of Byrd's experiences leading to his study are from Randolph C. Byrd, interviewed by John Sherrill, "The Therapeutic Effects of Intercessory Prayer," *Journal of Christian Nursing* 1 (1995): 21–23.

23. Byrd, "Intercessory Prayer," 21–23.

24. William Nolen, quoted in Wolinsky, "Prayers Do Aid Sick," 30.

25. C. Mann, "Meta-Analysis in the Breech," *Science* 249 (1990): 476–80.

26. E. Haraldsson and T. Thorsteinsson, "Psychokinetic Effects on Yeast: An Exploration Experiment," in *Research in Parapsychology 1972,* W. E. Roll, R. L. Morris, J. D. Morris, ed. (Metuchen, NJ: Scarecrow Press: 1973), 20–21.

27. The story of Minosch, the cat, is told by R. Scheib, *Utne Reader* (January–February 1996): 52–61. Bobby's story is studied by J. B. Rhine and S. R. Feather, "The Study of Cases of 'Psi-Trailing' in Animals," *Journal of Parapsychology* 26, no. 1 (1962): 1–21, and is told by Bill Schul, *The Psychic Power of Animals* (New York: Fawcett, 1977), 52.

28. A. H. Trapman, *The Dog, Man's Best Friend* (London: Hutchinson, 1929).

29. D. Bardens, *Psychic Animals* (New York: Barnes & Noble, 1996), 88.

30. Bardens, *Psychic Animals,* 14.

31. R. Sheldrake and P. Smart, "Psychic Pets: A Survey in North-West England," *Society for Psychical Research* 61 (1997): 353–64.

32. The following two paragraph quotes are from Rupert Sheldrake, *Seven Experiments That Could Change the World* (New York: Riverhead, 1995), 15–16.

33. Rupert Sheldrake, *A New Science of Life: The Hypothesis of Formative Causation,* 2d ed. (London: Anthony Blond, 1985).

34. Larry Dossey, *Recovering the Soul* (New York: Bantam, 1989), 1–11.

35. Charles Darwin's article was entitled "Origin of Certain Instincts," *Nature* 7 (1873): 417–18. Sheldrake has examined it and others in *Seven Experiments,* 33–72.

36. Sheldrake, *Seven Experiments,* 39.

37. Sheldrake, *Seven Experiments,* 52–53.

38. Sheldrake, *Seven Experiments,* 54–55.

39. Reported in the *London Daily Telegraph,* March 23, 1996; also, "Of All the Pigeon Lofts in All the World," *Fortean Times* 88 (July 1996): 10.

40. Loren Eiseley, cited in David Lorimer, *Whole in One* (London: Arkana, 1990), 72.

41. René Peoc'h, "Mise en évidence d'un effect psycho-physique chez l'homme et le poussin sur le tychoscope" (doctoral thesis, Univ. of Nantes, France, 1986).

42. René Peoc'h, "Psychokinetic Action of Young Chicks on the Path of an Illuminated Source," *Journal of Scientific Exploration* 9, no. 2 (1995): 223–29.

43. Dennis Gersten, M.D., letter to author, July 13, 1996. Gersten is author of *Are You Getting Enlightened Or Losing Your Mind? A Spiritual Program for Mental Fitness* (New York: Harmony, 1997).

44. Niels Bohr, quoted in J. A. Wheeler, "Not Consciousness but the Distinction Between the Probe and the Probed as Central to the Elemental Quantum Act of Observation," in *The Role of Consciousness in the Physical World,* ed. R. G. Jahn (Boulder: Westview Press, 1981), 94.

45. B. J. Dunne and R. G. Jahn, "Experiments in Remote Human/Machine Interaction," *Journal of Scientific Exploration* 6 (1992): 311.

46. Dunne and Jahn, "Remote Human/Machine Interaction," 311.

47. Robert G. Jahn and Brenda J. Dunne, "Penetration Parameters," *Margins of Reality* (New York: Harcourt Brace Jovanovich, 1987), 182–86. See also Larry Dossey, "Distance, Time, and Nonlocal Mind: Dare We Speak of the Implications?" *Journal of Scientific Exploration* 10, no. 3 (1996): 401–9.

48. B. J. Dunne, *Co-operator Experiments with an REG Device,* PEAR Technical Note 91005, Princeton Engineering Anomalies Research (Princeton: Princeton Univ. School of Engineering/Applied Science, 1991).

49. R. D. Nelson, G. J. Bradish, and Y. H. Dobyns, *The Portable PEAR REG: Hardware and Software Documentation,* Internal Document #92–1, Princeton Engineering Anomalies Research (Princeton: Princeton Univ. School of Engineering/Applied Science, 1992).

50. E. B. Titchener, "The Feeling of Being Stared At," *Science New Series* 8 (1898): 895–97; J. E. Coover, "The Feeling of Being Stared At," *American Journal of Psychology* 24 (1913): 570–75; J. J. Poortman, "The Feeling of Being Stared At," *Journal of the Society for Psychical Research* 40 (1959): 4–12.

51. L. Williams, "The Feeling of Being Stared At: A Parapsychological Investigation" (B.A. thesis, Department of Psychology, Univ. of Adelaide, South Australia, 1983). An abstract of this work was published in the *Journal of Parapsychology* 47 (1983): 59.

52. W. Braud, D. Schafer, and S. Andrews, "Electrodermal Correlates of Remote Attention: Autonomic Reactions to an Unseen Gaze," *Proceedings of the 33rd Annual Convention of the Parapsychological Association* (1990), 14–28; W. G. Braud, "Human Interconnectedness: Research Indications," *ReVision* 14, no. 3 (1992): 140–48; Marilyn J. Schlitz and Stephen LaBerge, "Covert Observation Increases Skin Conductance in Subjects Unaware of When They Are Being Observed: A Replication," *Journal of Parapsychology* 61 (1997): 185–96.

53. Richard Wiseman and Marilyn Schlitz, "Experimenter Effects and the Remote Detection of Staring," *Journal of Parapsychology* 61 (1997): 197–208.

54. Dennett's view is reported by Jim Holt, "Science Resurrects God," *Wall Street Journal,* December 24, 1997. Francis Crick's view is found in *The Astonishing Hypothesis: The Scientific Search for the Soul* (New York: Touchstone, 1995), 1.

55. Albert Einstein, quotation from the Internet at www.einstein.com, June 9, 1998; Niels Bohr, quoted in Werner Heisenberg, *Physics and Beyond,* trans. A. J. Pomerans (New York: Harper & Row, 1971), 114–15.

56. Eugene P. Wigner, "Are We Machines?" *Proceedings of the American Philosophical Society* 113, no. 2 (1969): 95–101.

57. Eugene Wigner, "Extension of the Area of Science," in *The Role of Consciousness in the Physical World,* ed. Robert G. Jahn, *AAAS Selected Symposium* 57 (Boulder: Westview, 1981), 13–14.

58. David Darling, "First World," *Omni* 17, no. 9 (1995): 4. See also *Soul Search* (New York: Villard Books, 1995).

59. George Wald, quoted in *Bulletin of the Foundation for Mind-Being Research* (September 1988): 3.

60. Gerald Feinberg, "Precognition—A Memory of Things Future," in *Quantum Physics and Parapsychology,* ed. L. Oteri (New York: Parapsychology Foundation, 1975), 54–73; Paul E. Meel and Michael Scriven, "Compatibility of Science and ESP," *Science* 123 (1956): 14–15; O. Costa de Beauregard, "The Paranormal Is Not Excluded from Physics," *Journal of Scientific Exploration* 12, no. 2 (1998): 315–20.

61. Henry Margenau, quoted in Lawrence LeShan, *The Science of the Paranormal* (Wellingborough, Northamptonshire, England: Aquarian Press, 1987), 118–19.

62. Dean I. Radin, Janine M. Rebman, and Maikwe P. Cross, "Anomalous Organization of Random Events by Group Consciousness: Two Exploratory Experiments," *Journal of Scientific Exploration* 10, no. 1 (1996): 143–68.

CHAPTER 3

1. Marilyn Ferguson, *The Aquarian Conspiracy* (New York: J. P. Tarcher, 1980).

2. Marilyn Ferguson, *Brain/Mind Bulletin* (June 1987): 2.

3. Andrew Greeley, "The Impossible: It's Happening," *Noetic Sciences Review* (Spring 1987): 7–9.

4. Greeley, "Impossible," 7–9.

5. Emily Wright Jaeger, *Christian Science Sentinel* 94, no. 50 (December 14, 1992): 17–21.

6. For this account I have relied on the insights of Rhea A. White, for which I am grateful: Rhea A. White, ed., *Special Issue, Exceptional Human Experience* 15, no. 1 (1997): 125–27.

7. Bill Russell, *Second Wind* (New York: Random House, 1979), 51–52, 60.

8. Russell, *Second Wind,* 66, 67, 68, 70.

9. Russell, *Second Wind,* 156–57.

10. Mona Lisa Schulz, *Awakening Intuition* (New York: Harmony, 1998), 31. Information on Salk may be found in Jonas Salk, *Anatomy of Reality: Merging Intuition and Reason* (New York: Columbia Univ. Press, 1983).

11. Ian Stevenson, *Telepathic Impressions* (Charlottesville: Univ. Press of Virginia, 1970), 5ff.

12. Schulz, *Awakening Intuition,* 106–7.

13. Schulz, *Awakening Intuition,* 107–8.

14. Schulz, *Awakening Intuition,* 108.

15. Leonard Michaels, in Naomi Epel, *Writers Dreaming* (New York: Carol Southern Books, 1993), 154–55.

16. Montague Ullman and Stanley Krippner, with Alan Vaughan, *Dream Telepathy: Experiments in Nocturnal ESP,* 2d ed. (Jefferson, NC: McFarland, 1989), 111–12. See also Stanley Krippner and Patrick Welch, *Spiritual Dimensions of Healing* (New York: Irvington, 1992), 188–89.

17. Robert L. Van de Castle, *Our Dreaming Mind* (New York: Ballantine, 1994), xxiii, 436–38.

18. Stanley Krippner, "Group Dreamworking in Berkeley," *The Inner Edge* (June/July 1998): 11–13.

19. Dean Radin, *The Conscious Universe* (San Francisco: HarperSanFrancisco, 1997), 296.

20. Russell Targ and Jane Katra, *Miracles of Mind: Exploring Nonlocal Consciousness and Spiritual Healing* (Novato, CA: New World Library, 1998), 81–82.

21. Van de Castle, *Our Dreaming Mind,* xxiii, 30–31. Van de Castle cites *Dreams and Dreaming* (Alexandria, VA: Time-Life Books, 1990), 30, 54–55.

22. Jessica Utts, "An Assessment of the Evidence for Psychic Functioning," *Journal of Scientific Exploration* 10, no. 1 (1996): 3–30.

23. H. E. Puthoff, "CIA-Initiated Remote Viewing Program at Stanford Research Institute," *Journal of Scientific Exploration* 10, no. 1 (1996): 63–76.

24. Puthoff, "CIA-Initiated Remote Viewing," 75.

25. Russell Targ, "Remote Viewing at Stanford Research Institute in the 1970s: A Memoir," *Journal of Scientific Exploration* 10, no. 1 (1996): 77–88. See also Targ and Katra, *Miracles of Mind.*

26. Radin, *Conscious Universe,* 292.

27. Ray Hyman, "Evaluation of Anomalous Phenomena," *Journal of Scientific Exploration* 10, no. 1 (1996): 31–58.

28. Puthoff, "CIA-Initiated Remote Viewing," 63–76.

29. Puthoff, "CIA-Initiated Remote Viewing," 77.

30. Arthur Koestler, "The Benefits of Impersonation," *The Act of Creation* (New York: Dell, 1964), 187–91.

31. Koestler, *Act of Creation,* 189.

32. Turtle Studios, 1203 King Street, Alexandria, VA.

33. William D. Rowe, "Physical Measurement of Episodes of Focused Group Energy," *Journal of Scientific Exploration* 12, no. 4 (1998): 569–81.

34. R. D. Nelson et al., "FieldRNG Anomalies in Group Situations," *Journal of Scientific Exploration* 10, no. 10 (1996): 111–41.

35. Paul Ekman, "Expression and the Nature of Emotion," in *Approaches to Emotion,* ed. K. Scherer and P. Ekman (Hillsdale, NJ: Lawrence Erlbaum, 1984).

36. Nicholas R. S. Hall, Maureen O'Grady, and Denis Calandra, "Transformation of Personality and the Immune System," *Advances* 10, no. 4 (1994): 7–15.

37. Bernie S. Siegel, letter to the editor, *Advances* 11, no. 2 (1995): 2.

38. John Ruskin, quoted in D. Bardens, *Psychic Animals* (New York: Barnes & Noble, 1996), 79.

39. J. Arehart-Treichel, "Pets: The Health Benefits," *Science News* 121 (1982): 220.

40. K. MacInnis, "Blackie," *American Journal of Nursing* 91, no. 7 (1991): 84.

CHAPTER 4

1. J. A. Henslee et al., "The Impact of Premonitions of SIDS on Grieving and Healing," *Pediatric Pulmonology* 16 (1993): 393.

2. Richard Smith, "Where Is the Wisdom . . . ?" *British Medical Journal* 303 (1991): 798–99.

3. Names and places have been changed to preserve anonymity. May 1997.

4. Mona Lisa Schulz, *Awakening Intuition* (New York: Harmony, 1998), 47.

5. D. Schneider, *Revolution in the Body-Mind: Forewarning Cancer Dreams and the Bioplasma Concept* (East Hampton, NY: Alexa Press, 1976); L. H. Bartmeier, "Illness Following Dreams," *International Journal of Psychoanalysis* 31 (1950): 8.

6. V. N. Kasatkin, *Teoriya Snovidenii (Theory of Dreams)* (Leningrad: Meditsina, 1967), 352, cited in R. C. Van de Castle, *Our Dreaming Mind* (New York: Ballantine, 1994), 367.

7. E. G. Mitchell, "The Physiological Diagnostic Dream," *New York Medical Journal and Medical Record* 118 (1923): 416–17.

8. Barbara Montgomery Dossey, *Florence Nightingale: Mystic, Visionary, Healer* (Springhouse, PA: Springhouse Corporation, 1999).

9. William Dement, *Some Must Watch While Some Must Sleep* (San Francisco: W. H. Freeman, 1972), 102.

10. Schulz, *Awakening Intuition,* 48.

11. Bernard R. Grad, "Some Biological Effects of Laying On of Hands: A Review of Experiments with Animals and Plants," *Journal of the American Society of Psychical Research* 51a (1965): 95–127.

12. Letter to author, June 1998. Names and location have been changed to protect anonymity.

13. Related to Larry Dossey by Barbara Dossey, July 1998. Name is changed to protect anonymity.

14. Letter to author, January 1996. Names and location have been changed to protect anonymity.

15. Elaine Haire, letter to author, January 1997

16. Harold McKinney, letter to author, January 1997.

17. Irene Spencer, letter to author, April 20, 1998.

18. Isabel Allende, in Naomi Epel, *Writers Dreaming* (New York: Carol Southern Books, 1993), 19–20.

19. Names and locations have been changed to protect anonymity. Date of letter is July 1995.

20. A. N. Exton-Smith and M. D. Cantaub, "Terminal Illness in the Aged," *Lancet* 2 (1961): 305.

21. Connie Hernandez, M.D., letter to author, October 1996.

22. Earlier we saw several examples of telesomatic events that have been researched by University of Virginia psychiatrist Ian Stevenson [pp. 101–102].

23. Linda Whitson, letter to author, August 1999.

24. Larry Dossey, *Recovering the Soul* (New York: Bantam, 1989).

25. Sir Laurens van der Post, I later discovered, has explored the role of the praying mantis in premodern cultures as a messenger and bringer of wisdom. See his book *Mantis Carol* (New York: Island Press, 1994).

26. Barbara Mathews, letter to author, January 27, 1999. Web site for Multisensory Learning: www.interlog.com/~multisen/.

27. Larry Dossey, "Cancelled Funerals: A Look at Miracle Cures," *Alternative Therapies* 4, no. 2 (1998): 10–18, 116–20.

28. Caryle Hirshberg and Marc Ian Barasch, *Remarkable Recovery* (New York: Riverhead Books, 1995), 117–23.

29. Kat Duff, *The Alchemy of Illness* (New York: Pantheon, 1993), xi–xiv, 13.

30. Hirshberg and Barasch, *Remarkable Recovery,* 37–38.

31. Milton Sacks, quoted in the *Washington Post,* April 3, 1994.

32. H. J. G. Bloom, W. W. Richardson, and E. J. Harries, "Natural History of Untreated Breast Cancer (1805–1933)," *British Medical Journal* 2 (1962): 213–21.

33. Hirshberg and Barasch, *Remarkable Recovery,* 14–15.

34. Hirshberg and Barasch, *Remarkable Recovery,* 137.

35. R. C. S. Ayres, "Spontaneous Regression of Hepatocellular Carcinoma," *Gut* 3, no. 6 (1990): 722–24.

36. Hirshberg and Barasch, *Remarkable Recovery,* 332–33.

37. J. Achterberg and G. F. Lawlis, "Letters: Human Research and Studying Psychosocial Interventions for Cancer," *Advances* 8, no. 4 (1992): 4.

38. Lewis Thomas, quoted in P. Norris, "Self-Regulation Through Imagery," part 2, "Visualization & Cancer Therapy," *Newsletter of the International Society for the Study of Subtle Energy & Energy Medicine* 3, no. 2 (1992): 2.

39. Duff, *Alchemy of Illness,* 13.

40. Rhea A. White, "Kat Duff: From Healing Self to World Renewal," *Exceptional Human Experience: Special Issue* 15, no. 1 (1997): 103–6.

41. Duff, *Alchemy of Illness,* 88–89.

42. Duff, *Alchemy of Illness,* 89.

43. Duff, *Alchemy of Illness,* 92–93.

44. Duff, *Alchemy of Illness,* 89, 141.

45. Lyall Watson, *The Nature of Things: The Secret Life of Inanimate Objects* (Rochester, VT: Destiny Books, 1992), 215.

46. Watson, *Nature of Things,* 199–200.

47. Watson, *Nature of Things,* 194.

48. R. L. Morris, "Applied Psi in the Context of Human-Equipment Interaction Systems," *Proceedings of Symposium on Applications of Anomalous Phenomena* (1983), cited in Watson, *Nature of Things,* 195.

49. *Celebrity* (England), November 5, 1987, cited in Watson, *Nature of Things,* 195.

50. Edward Tenner, *Why Things Bite Back: Technology and the Revenge of Unintended Consequences* (New York: Alfred A. Knopf, 1996), 4–5.

51. *Reader's Digest,* August 1982.

52. Michael Shallis, *The Electric Shock Book* (London: Souvenir, 1988), cited in Watson, *Nature of Things,* 194.

53. Gallup poll reported in *Life* magazine, March 1994.

54. Larry Dossey, "Scientific Studies," *Be Careful What You Pray For* (San Francisco: HarperSanFrancisco, 1997), 165–92.

55. Glen Rein, *Quantum Biology: Healing with Subtle Energy* (Palo Alto: Quantum Biology Research Labs, 1992). See also Daniel J. Benor, *Healing Research,* vol. 1 (Munich: Springer Verlag, 1993), 138–42; and

Glen Rein, "The Scientific Basis for Healing with Subtle Energies," appendix A in *Healing with Love,* by Leonard Laskow (San Francisco: HarperSanFrancisco, 1992), 279–319.

56. David J. Muehsam et al., "Effect of Qigong on Cell-Free Myosin Phosphorylation: Preliminary Experiments," *Subtle Energies* 5, no. 1 (1994): 93–108.

57. William Braud, Gary Davis, and Robert Wood, "Experiments with Matthew Manning," *Journal of the Society for Psychical Research* 50 (1979): 199–223.

58. Jean Barry, "General and Comparative Study of the Psychokinetic Effect on a Fungus Culture," *Journal of Parapsychology* 32, no. 4 (1968): 237–43.

59. William H. Tedder and Melissa L. Monty, "Exploration of Long-Distance PK: A Conceptual Replication of the Influence on a Biological System," in *Research in Parapsychology 1980,* ed. W. G. Roll et al. (Metuchen, NJ: Scarecrow Press, 1981), 90–93.

60. Carroll B. Nash, "Psychokinetic Control of Bacterial Growth," *Journal of the Society for Psychical Research* 51 (1982): 217–21.

61. On the psychospiritual immune system, see Dossey, "Protection," in *Be Careful,* 8. On protecting ourselves, see 195–217.

62. St. Teresa, letter to Don Lorenzo de Cepeda, January 1577, cited in Jean Lanier, "From Having a Mystical Experience to Becoming a Mystic—Reprint and Epilogue," *ReVision* 12, no. 1 (Summer 1989): 41–44. On Suzuki Roshi, see Helen Tworkov, *Zen in America* (San Francisco: North Point Press, 1989), 225.

63. Charles T. Tart, "Fears of the Paranormal in Ourselves and Our Colleagues: Recognizing Them, Dealing with Them," *Subtle Energies* 5, no. 1 (1994): 35–67.

64. Brother David Steindl-Rast, "Learning to Die," *Parabola* 2, no. 1 (1977): 22–31.

CHAPTER 5

1. Nancy Tomes, *The Gospel of Germs: Men, Women, and the Microbe in American Life* (Cambridge: Harvard Univ. Press, 1998), 5.

2. Tomes, *Gospel of Germs,* 24–25.

3. Tomes, *Gospel of Germs,* 25.

4. Tomes, *Gospel of Germs,* 25.

5. Jeffrey A. Fisher, *The Plague Makers* (New York: Simon & Schuster, 1994), 17–18.

6. For the Type A personality, see M. Friedman and R. H. Rosenman, *Type A Behavior and Your Heart* (New York: Alfred A. Knopf, 1974). For emotional states after heart attacks, see N. H. Cassem, T. P. Hackett, and H. A. Wishnie, "The Coronary Care Unit: An Appraisal of Its Psychological Hazards," *New England Journal of Medicine* 279 (1968): 13–65. For the influence of social contact in resistance to disease, see L. F. Berman and S. L. Syme, "Social Networks, Host Resistance, and Mortality: A Nine-Year Follow-Up Study of Alameda County Residents," *American Journal of Epidemiology* 109 (1979): 186–204. On the correlation of stress with health problems, see T. H. Holmes and R. H. Rahe, "The Social Readjustment Rating Scale," *Journal of Psychosomatic Research* 11 (1967): 213–18. On job dissatisfaction as predictor of heart attacks, see *Work in America: Report of a Special Task Force to the Secretary of Health, Education, and Welfare* (Cambridge: MIT Press, 1973). On spousal relationship and angina, see J. H. Medalie and U. Goldbourt, "Angina Pectoris Among 10,000 Men II: Psychosocial and Other Risk Factors as Evidenced by a Multivariate Analysis of Five-Year Incidence Study," *American Journal of Medicine* 60 (1976): 910–21. On relationships with parents as predictors of cancer, see C. B. Thomas, "Precursors of Premature Disease and Death: The Predictive Potential of Habits and Family Attitudes," *Annals of Internal Medicine* 85 (1976): 653–58. On weakened immune

systems during bereavement, see S. J. Schleifer, "Bereavement and Lymphocyte Function" (paper presented to the annual meeting of the American Psychiatric Association, San Francisco, May 1980). On the benefits of relaxed states, see Herbert Benson, *The Relaxation Response* (New York: William Morrow, 1975).

7. On imagery and visualization techniques, see O. Carl Simonton, Stephanie Matthews-Simonton, and James Creighton, *Getting Well Again* (Los Angeles: J. P. Tarcher, 1978), also Jeanne Achterberg and G. Frank Lawlis, *Imagery of Cancer* (Champaign, IL: Institute for Personality and Ability Testing, 1978). On using meditation to lower blood pressure and cholesterol levels, see M. Cooper and M. Aygen, "Effect of Meditation on Blood Cholesterol and Blood Pressure," *Journal of the Israel Medical Association* 95 (1978): 1; also R. A. Stone and J. DeLeo, "Psychotherapeutic Control of Hypertension," *New England Journal of Medicine* 294 (1976): 80.

8. F. H. Garrison, *History of Medicine,* 4th ed. (Philadelphia: W. B. Saunders, 1928), 414, 619–20, 757.

9. Alan Gauld, *A History of Hypnotism* (New York: Cambridge Univ. Press, 1992), 39–52.

10. Brian Inglis, *Natural and Supernatural: A History of the Paranormal* (Bridport, Dorset, England: Prism Press, 1992), 146, 149, 152–53; Gauld, *History of Hypnotism,* 150–52.

11. Inglis, *Natural and Supernatural,* 157.

12. Inglis, *Natural and Supernatural,* 255–56.

13. Gauld, *History of Hypnotism,* 108–9.

14. Inglis, *Natural and Supernatural,* 211–14.

15. Norman Shealy and Carolyn Myss, *The Creation of Health* (Walpole, NH: Stillpoint, 1988).

16. Judith Orloff, *Second Sight* (New York: Time Warner, 1996), 359–61.

17. Mona Lisa Schulz, *Awakening Intuition* (New York: Harmony, 1998), 2.

18. Schulz, *Awakening Intuition,* 9.

19. Orloff, *Second Sight,* xvii–xviii.

20. Ellen Idler and Stanislav Kasl, "Health Perceptions and Survival: Do Global Evaluations of Health Status Really Predict Mortality?" *Journal of Gerontology* 46, no. 2 (1991): S55–65.

21. Larry VandeCreek, Elizabeth Rogers, and Joanne Lester, "Use of Alternative Therapies Among Breast Cancer Outpatients Compared with the General Population," *Alternative Therapies* 5, no. 1 (1999): 71–76; John A. Astin, "Why Patients Use Alternative Medicine—Results of a National Survey," *Journal of the American Medical Association* 279, no. 19 (1998): 1548–53; D. M. Eisenberg et al., "Unconventional Medicine in the United States: Prevalence, Costs, and Patterns of Use," *New England Journal of Medicine* 328 (1993): 246–52.

22. Claire Sylvia, *A Change of Heart: A Memoir,* with William Novak (Boston: Little Brown, 1997).

23. Peter Fenwick and Elizabeth Fenwick, "Transmission Theories [of Consciousness]," *The Truth in the Light* (New York: Berkley Books, 1997), 259–62.

24. On nonlocal mind affecting humans and human tissue, see Glen Rein, "A Psychokinetic Effect on Neurotransmitter Metabolism: Alterations in the Degradative Enzyme Monoamine Oxidase," in *Research in Parapsychology 1985,* ed. Debra H. Weiner and Dean Radin (Metuchen, NJ: Scarecrow Press, 1986), 77–80, also W. G. Braud, "Distant Mental Influence of Rate of Hemolysis of Human Red Blood Cells," *Journal of the American Society for Psychical Research* 84 (1990): 1–24. On its effects on animals, see Graham K. Watkins and Anita M. Watkins, "Possible PK Influence on the Resuscitation of Anesthetized Mice," *Journal of Parapsychology* 35, no. 4 (1971): 257–72; Graham K.

Watkins, Anita M. Watkins, and Roger A. Wells, "Further Studies on the Resuscitation of Anesthetized Mice," in *Research in Parapsychology 1972,* ed. W. C. Roll et al. (Metuchen, NJ: Scarecrow Press, 1973), 157–59; R. Wells and G. Watkins, "Linger Effects in Several PK Experiments," *Research in Parapsychology 1974,* ed. W. C. Roll et al. (Metuchen, NJ: Scarecrow Press, 1975), 143–47; B. R. Grad, "Some Biological Effects of Laying-On of Hands: A Review of Experiments with Animals and Plants," *Journal of the American Society for Psychical Research* 59a (1965): 95–127; B. Grad, R. J. Cadoret, and G. I. Paul, "The Influence of an Unorthodox Method of Treatment of Wound Healing in Mice," *International Journal of Parapsychology* 3 (1961): 5–24. On the effects of nonlocal mind on plants, see Grad, "Some Biological Effects," 124–25. For its effects on microorganisms, see Jean Barry, "General and Comparative Study of the Psychokinetic Effect on a Fungus Culture," *Journal of Parapsychology* 32 (1968): 237–43; William H. Tedder and Melissa L. Monty, "Exploration of Long-Distance PK: A Conceptual Replication of the Influence on a Biological System," in *Research in Parapsychology 1980,* ed. W. B. Roll et al. (Metuchen, NJ: Scarecrow Press, 1981), 90–93; Carroll B. Nash, "Psychokinetic Control of Bacterial Growth," *Journal of the American Society for Psychical Research* 51 (1982): 217–21; Carroll B. Nash, "Test of Psychokinetic Control of Bacterial Mutation," *Journal of the American Society for Psychical Research* 78, no. 2 (1984): 145–52; E. Haraldsson and T. Thorsteinsson, "Psychokinetic Effects on Yeast: An Exploratory Experiment," in *Research in Parapsychology 1972,* ed. W. C. Roll et al. (Metuchen, NJ: Scarecrow Press, 1973), 20–21; and J. Solfvin, "Mental Healing," in *Advances in Parapsychological Research,* ed. Stanley Krippner (Jefferson, NC: McFarland, 1984), 31–63. For summaries of these studies, see Daniel Benor, *Healing Research,* vol. 1 (Munich: Helix Verlag, 1993); Larry Dossey, *Healing Words* (San Francisco: Harper-SanFrancisco, 1993).

25. Rupert Sheldrake, "Experimenter Effects in Scientific Research: How Widely Are They Neglected?" *Journal of Scientific Exploration* 12, no. 1 (1998), 73–78.

26. Fred Sicher et al., "A Randomized Double-Blind Study of the Effect of Distant Healing in a Population with Advanced AIDS: Report of a Small-Scale Study," *Western Journal of Medicine* 169, no. 6 (1998): 356–63.

27. Daniel P. Wirth, "Unorthodox Healing: The Effect of Therapeutic Touch on the Healing Rate of Full Thickness Dermal Wounds," *Subtle Energies* 1, no. 1 (1990): 1–20.

28. Grad, "Some Biological Effects," 95–127.

29. Braud, "Distant Mental Influence," 1–24.

30. Dean Radin, "Mental Interactions with Living Organisms," in *The Conscious Universe* (San Francisco: HarperSanFrancisco, 1997), 147–56; Barry, "Fungus Culture," 237–43; Tedder and Monty, "Long-Distance PK," 90–93; Nash, "Bacterial Growth," 217–21; Nash, "Bacterial Mutation," 145–52; Haraldsson and Thorsteinsson, "Psychokinetic Effects on Yeast," 20–21.

31. For the controlled studies, see Robert G. Jahn and Brenda J. Dunne, "Precognitive Remote Perception," in *Margins of Reality: The Role of Consciousness in the Physical World* (New York: Harcourt Brace Jovanovich, 1987), 149–91, and Dean Radin, "Telepathy" and "Perception at a Distance," in *The Conscious Universe* (San Francisco: HarperSanFrancisco, 1997), 61–90, 91–110. For case reports, see Ian Stevenson, *Telepathic Impressions: A Review and Report of 35 New Cases* (Charlottesville: Univ. Press of Virginia, 1970).

32. G. K. Watkins and A. M. Watkins, "Possible PK Influence," 257–72; G. Watkins, A. Watkins, and R. Wells, "Further Studies," 157–59; R. Wells and G. Watkins, "Linger Effects," 143–47.

33. William G. Braud and Marilyn J. Schlitz, "Psychokinetic Influence on Electrodermal Activity," *Journal of Parapsychology* 47 (1983): 95–119.

34. Larry Dossey, "'Let It Be' or 'Make It Happen'? The Spindrift Studies," in *Healing Words*, 97–108; Dossey, *Be Careful What You Pray*

For (San Francisco: HarperSanFrancisco, 1997), 144–45, 173–75. On personality types, see Dossey, "From Personality Types to Spiritual Types," in *Healing Words*, 94–101.

35. Jeanne Achterberg, *Woman as Healer* (Boston: Shambhala, 1990).

36. William B. Stewart and Meg Page, "Bearing Witness to the Evolution of Contemporary Medicine: The Institute for Health and Healing at California Pacific Medical Center, San Francisco," *Healthcare Forum Journal* (November/December 1998): 38–40.

37. Kenneth R. Pelletier, "Life with the New Roommate: Alternative Medicine Moves In with Conventional Medicine," *Healthcare Forum Journal* (November/December 1998): 35–37, 41.

38. Wayne Jonas, "Alternative Medicine and the Conventional Practitioner," *Journal of the American Medical Association* 279, no. 9 (1998): 708–9. See also A. Furnham and J. Forey, "The Attitudes, Behaviors, and Beliefs of Patients of Conventional vs. Complementary (Alternative) Medicine," *Journal of Clinical Psychology* 50 (1994): 458–69; C. Vincent, A. Furnham, and M. Willsmore, "The Perceived Efficacy of Complementary and Orthodox Medicine in Complementary and General Practice Patients," *Health Education Theory and Practice* 10 (1995): 395–405; and D. M. Eisenberg, "The Invisible Mainstream," *Harvard Medical Alumni Bulletin* (1996): 20–25.

39. On physicians "staying out of the God business," see R. P. Sloan, E. Bagiella, and T. Powell, "Religion, Spirituality, and Medicine," *Lancet* 353 (1999): 664–67. For a response to this article, see Larry Dossey, "Do Religion and Spirituality Matter in Health? A Response to the Recent Article in *The Lancet*," *Alternative Therapies* 5, no. 3 (1999): 16–18. Post is quoted in Mike Mitka, "Getting Religion Seen as Help in Being Well," *Journal of the American Medical Association* 280, no. 22 (1998): 1896–97.

40. The AAPC's Code of Ethics may be obtained from the American Association of Pastoral Counselors, 9504A Lee Highway, Fairfax, VA 22031-2303; Internet: info@aapc.org. Dr. Thomas Oxman's

results are found in T. E. Oxman, D. H. Freeman, and E. D. Manheimer, "Lack of Social Participation or Religious Strength or Comfort as Risk Factors after Cardiac Surgery in the Elderly," *Psychosomatic Medicine* 57 (1995): 5–15. For the survey of hospital patients, see D. E. King and B. Bushwick, "Beliefs and Attitudes of Hospital Inpatients About Faith Healing and Prayer," *Journal of Family Practice* 39 (1994): 349–52.

41. Letter to author, February 1995. Names and locations have been changed to preserve anonymity.

42. Andrew Weil, quoted in *Wise Words,* ed. Mary Buckley (Carlsbad, CA: Hay House, 1998). See also Weil's *Spontaneous Healing* (New York: Alfred A. Knopf, 1995).

43. For statistics on medical schools, see Jeffrey S. Levin, David B. Larson, and Christina M. Puchalski, "Religion and Spirituality in Medicine: Research and Education," *Journal of the American Medical Association* 278, no. 9 (1997): 792–93. These authors list approximately thirty medical schools that feature courses along these lines. Since then the number has increased substantially. Studies published in the *JAMA* include Charles Marwick, "Should Physicians Prescribe Prayer for Health? Spiritual Aspects of Well-Being Considered," *Journal of the American Medical Association* 273, no. 20 (1995): 1561–62; Mike Mitka, "Getting Religion Seen as Help in Being Well," *Journal of the American Medical Association* 280, no. 22 (1998): 1896–97; and Levin, Larson, and Puchalski, "Religion and Spirituality," 792–93. *Alternative Therapies in Health and Medicine* is published by Innovision Communications, 101 Columbia, Aliso Viejo, CA 92656; phone 800-899-1712; Internet: www.alternative-therapies.com.

44. Mitka, "Getting Religion," 1896–97.

45. Larry Dossey, *Meaning and Medicine* (New York: Bantam, 1991), 209.

46. Pupul Jayakar, *Krishnamurti: A Biography* (San Francisco: Harper & Row, 1986), 485–86.

47. E. J. Cassell, quoted by C. Laine and F. Davidoff, "Patient-Centered Medicine," *Journal of the American Medical Association* 275 (1996): 152–56.

48. Laine and Davidoff, "Patient-Centered Medicine," 152–56.

CHAPTER 6

1. J. Lazarou, B. H. Pomeranz, and P. N. Corey, "Incidence of Adverse Drug Reactions in Hospitalized Patients," *Journal of the American Medical Association* 279 (1998): 1200–1205.

2. Rodney H. Falk, "The Death of Death with Dignity," *American Journal of Medicine* 77, no. 5 (1984): 775–76; T. L. Petty, "Mechanical Last 'Rights,'" *Archives of Internal Medicine* 142 (1982): 1442–43.

3. F. Stenn, "Thoughts of a Dying Physician," *Forum on Medicine* 3 (1980): 718–19.

4. Philippe Aries, *Western Attitudes Toward Death* (Baltimore and London: Johns Hopkins Univ. Press, 1974), 1–25; Patrick B. Friel, "Death and Dying," *Annals of Internal Medicine* 97 (1982): 767–71.

5. Friel, "Death and Dying," 767–71.

6. Friel, "Death and Dying," 767–71.

7. G. Gorer, *Death, Grief and Mourning* (New York: Anchor/Doubleday, 1967), 192–99.

8. C. G. Jung, *Memories, Dreams, Reflections* (New York: Vintage, 1965), 314.

9. Virginia H. Hine, "Holistic Dying: The Role of the Nurse Clinician," *Topics in Clinical Nursing* 3, no. 4 (1982): 45–54.

10. Virginia H. Hine, *Last Letter to the Pebble People* (Santa Cruz, CA: Unity Press, 1979).

11. Karlis Osis and Erlendure Haraldsson, *What They Saw . . . At the Hour of Death* (Norwalk, CT: Hastings House, 1997).

12. Kathleen Dowling Singh, *The Grace in Dying: How We Are Transformed Spiritually as We Die* (San Francisco: HarperSanFrancisco, 1998).

13. Kathleen Singh, letter to the author, January 1996.

14. Timothy Ferris, "A Cosmological Event," *New York Times Magazine,* December 15, 1991, 44–52.

15. Sogyal Rinpoche, quoted in *Wise Words,* ed. Mary Buckley (Carlsbad, CA: Hay House, 1998).

16. Ferris, "Cosmological Event," 52.

17. Annalee R. Oakes, "Near-Death Events and Critical Care Nursing," *Topics in Clinical Nursing* 3, no. 3 (1981): 61–78.

18. Oakes, "Near-Death Events," 69.

19. Dawna Markova, *No Enemies Within* (Berkeley: Conari Press, 1994).

20. C. M. Saunders, "The Challenge of Terminal Care," in *Scientific Foundations of Oncology,* ed. T. Symington and R. L. Carter (London: William Heineman Medical Books, 1975), 673–78.

21. Raymond Moody, *Life After Life* (Covington, GA: Mockingbird Books, 1975). On statistics of near-death experiences, see Kenneth Ring, *Lessons from the Light* (New York: Plenum/Insight, 1998), 212.

22. August L. Reader III, "The Internal Mystery Plays: The Role and Physiology of the Visual System in Contemplative Practices," *Alternative Therapies* 1, no. 4 (1995): 54–63. All quotes from Reader below refer to this article.

23. Dannion Brinkley, *At Peace in the Light* (New York: HarperCollins, 1995).

24. Arthur S. Berger, "Quoth the Raven: Bereavement and the Paranormal," *Omega* 31 (1995): 1–10. See also A. S. Berger, *Evidence for Life After Death: A Casebook for the Toughminded* (Springfield, IL: Charles C. Thomas, 1988). Many of the observations that follow are from Berger's article "Quoth the Raven," for which I am grateful.

25. Bill Guggenheim and Judy Guggenheim, *Hello from Heaven!* (New York: Bantam, 1995), 363–64.

26. W. D. Rees, "The Bereaved and Their Hallucinations," in *Bereavement and Its Psychosocial Aspects,* ed. B. Schoenberg et al. (New York: Columbia Univ. Press, 1975).

27. G. Ginsburg, *To Live Again* (Los Angeles, J. P. Tarcher, 1987), 47, reported in Berger, "Quoth the Raven," 2.

28. For the mortality of surviving spouses, see Larry Dossey, "Broken Hearts: The Toxicity of Bereavement," *Meaning & Medicine* (New York: Bantam, 1991), 88–96. For Schleifer's estimates, see Steven J. Schleifer et al., "Suppression of Lymphocyte Stimulation Following Bereavement," *Journal of the American Medical Association* 250, no. 3 (1983): 374–77.

29. Ian Stevenson, *Children Who Remember Previous Lives* (Charlottesville: Univ. Press of Virginia, 1987). The Voltaire quote is from "La princesse de Babylone," in *Romans et Contes* (Paris: Editions Garnier Frères, 1960), 366.

30. Robert Payne, *The Life and Death of Mahatma Gandhi* (New York: Konecky & Konecky, 1969), 372.

POSTSCRIPT

1. D. M. Eisenberg et al., "Trends in Alternative Medicine Use in the United States, 1990–1997," *Journal of the American Medical Association* 280, no. 18 (1998): 1569–75.

2. J. A. Astin, "Why Patients Use Alternative Medicine," *Journal of the American Medical Association* 279, no. 19 (1998): 1548–53.

INDEX